LAND of the STRIPED STALKER

Wildlife of Madhya Pradesh

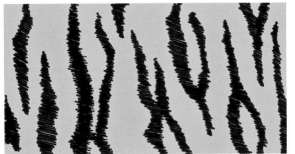

LAND of the STRIPED STALKER

Wildlife of Madhya Pradesh

Text
Rajesh Gopal

Photographs
Rajesh Bedi

The heart of
Incredible India
Madhya Pradesh Tourism

wisdom tree

First published in 2010 by
Wisdom Tree
4779/23
Ansari Road
Darya Ganj
New Delhi - 110002
Phone: 91 - 11 - 23247966/ 67/ 68
wisdomtreebooks@gmail.com

The project has been supported by Madhya Pradesh State Tourism Development Corporation Limited.

Printed in India.

ISBN 978-81-8328-153-9

Design: tiffinbox

Front cover: *A tiger on the prowl at Kanha Tiger Reserve.*
Back cover: *A herd of Spotted deer graze contentedly in the golden afternoon light at Kanha Tiger Reserve.*

CONTENTS

ACKNOWLEDGEMENTS

Several colleagues have helped me in the preparation of this book, and it may not be possible to list them all. I especially appreciate the encouragement and the help from Y.V. Jhala, Qamar Qureshi, Rajesh Bedi, Naresh Bedi, H.S. Negi, Aseem Srivastava, S.P. Yadav, Gopa Pandey, Sheetal Bisht, P.R. Sinha and Narendra Kumar.

◀ LEFT

The grasslands of Bandhavgarh Tiger Reserve are softly illuminated by the early morning light.

PREFACE

Madhya Pradesh has a diversity of life forms owing to its rich, unique geographical and climatic attributes. The large number of protected areas in the state highlight the concern and efforts of the state government to conserve and protect its floral and faunal heritage. The state is virtually a 'tigerland', and places like Kanha, Bandhavgarh, Pench, Satpura and the nearby forests are home to the bulk of the population of tigers, their co-predators, prey and habitat of Central India. I had the good fortune of working for over thirteen years in Kanha and Bandhavgarh. Every day was a learning experience for me during these wonderful years, which I would love to relive if given a chance!

This book is a modest effort to present a broad picture of the prominent protected areas, tiger reserves and flora and fauna of Madhya Pradesh. However, the thrill of seeing a tiger in the wild, or hearing the rutting call of a Barasingha stag, the alarm call of a Spotted deer or the urgent calls of a peafowl are those moments which should best be experienced. The direct sighting of a wild animal in its natural habitat is the best understanding of Nature.

The tiger is India's iconic national animal. It widely inhabits forested and non-forested natural habitats in seventeen states of India including Madhya Pradesh. Owing to its significant ecological position, the conservation status of the tiger in our ecosystems signifies the status of their health. The good health of these ecosystems in turn signifies that the quality of ecological services rendered by them, including ensuring water security – critical to our survival and to sustainable development – is optimum.

RIGHT

The rolling hills and verdant forested areas of Bandhavgarh Tiger Reserve provide ample cover for a variety of wildlife.

This is unfortunately not the case. The degradation and fragmentation of forests and non-forested natural areas across the length and breadth of our country reflects the decline in ecological services. This is seen in the form of poor water regime and the loss of our unique biodiversity. The decline has also undermined the productivity of the livestock of people inhabiting the forested regions, further aggravating their impoverishment.

The need for conserving the tiger, other wild animals and their habitats must therefore be seen in the light of these ecological imperatives. This calls for an integrated, holistic approach

towards tourism as well as managing land use. This uphill task can only be achieved by mainstreaming wildlife concerns in various types of land use operating in protected areas where the primary goal may not be wildlife conservation.

Over the years, protected areas in many states have witnessed a steep increase in the number of visitors, and Madhya Pradesh is no exception. This is indeed a welcome sign for fostering tourism, but is a cause for concern as well. We must not forget the fact that these protected areas and core areas of tiger reserves are the birth place of the tiger and other wild animals, requiring complete tranquility from any kind of disturbance. Thus, it becomes an ecological imperative to foster low key, ecologically sustainable tourism restricted to the fringes of these reserves and protected areas. The outer corridor connectivity of these areas should not be used for large-scale tourism infrastructure, since such actions are detrimental to tigers and other wild animals. The Government of India, state governments and various institutions of civil society are jointly responsible for the conservation of wildlife. The Constitution of India urges us to protect our wild denizens. Our visits to protected areas should be both for the enjoyment of our natural heritage and to elicit public support for its conservation.

It is earnestly hoped that the information provided in this book would not only help visitors in planning their trips to the protected areas and tiger reserves of Madhya Pradesh, but will also foster awareness on the ecological imperatives to protect our wildlife to give an assured future to our wild flora and fauna. A visitor to a protected area must assume the role of a naturalist so that the visit is rewarding without causing any distress to the natural inhabitants of our forests.

INTRODUCING
THE TIGER STATE

Madhya Pradesh is a large, landlocked state located in the heartland of India. It shares borders with other states like Gujarat, Rajasthan, Uttar Pradesh, Chhattisgarh and Maharashtra. In terms of area, the state ranks second in the country, with forest cover occupying almost one-third of its geographical area.

The geographical centre of India is situated at Karondi in the Jabalpur district of Madhya Pradesh. The physical features of the state are very interesting, comprising of plateaus, hills, plains and rivers.

The Tropic of Cancer is more or less parallel to Narmada River, which originates from Amarkantak in the Vindhya range of the state, and flows westwards. On the southern side of its valley, the Satpura Hills form a triangular elevation, serving as a watershed. Several rivers flow through the state, which include Chambal, Betwa, Shipra and Sone.

◂ LEFT

A tigress in royal repose with a cub in Kanha Tiger Reserve.

STATUS OF BIODIVERSITY

Madhya Pradesh is gifted with a rich biodiversity. It is not just confined to forests but also extends to reservoirs, wetlands, livestock, and agro-ecosystems. Broadly, the biodiversity of Madhya Pradesh can be categorised into forests (including national parks, sanctuaries and tiger reserves), grasslands and wetlands.

Tiger Reserves/ National Parks	
1.	Bandhavgarh TR
2.	Fossil NP
3.	Kanha TR
4.	Madhav NP
5.	Panna TR
6.	Pench TR
7.	Sanjay TR
8.	Satpura TR
9.	Van Vihar NP

Wildlife Sanctuaries	
1.	Bori
2.	Bagdara
3.	Phen
4.	Ghatigaon
5.	Gandhisagar
6.	Karera
7.	Ken Ghariyal
8.	Kheoni
9.	Narsinghgarh
10.	N. Chambal
11.	Nauradehi
12.	Pachmari
13.	Panpatha
14.	Kuno
15.	Pench
16.	Ratapani
17.	Sanjay Dubri
18.	Singhori
19.	Son Ghariyal
20.	Sardarpur
21.	Sailana
22.	Ralamandal
23.	Orchha
24.	Gangau
25.	Veeranga Durgawati

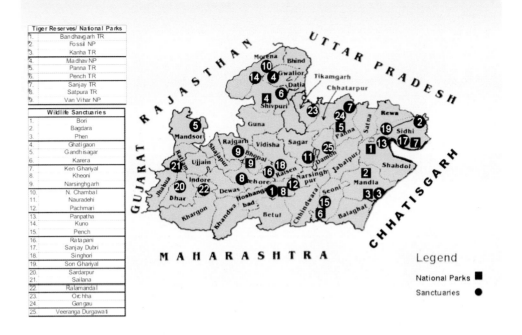

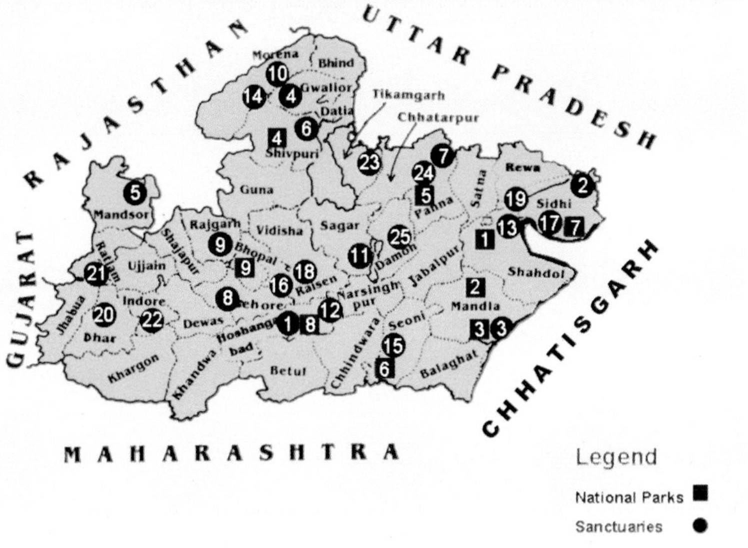

The state is home to 4,000 plant species, which includes 165 pteridophytes, around 50 bryophytes and 500 species of medicinal plants. The rare and endemic flora of the state includes different species of ferns including tree ferns, mule's-foot fern, moonworts and royal fern. There are around 22,000 cultivars of rice in the undivided Madhya Pradesh characterised by their distinct aroma (*Dilbuxa*, *Chinmouri*, *Tulsi* and *Amrat*) or resistance to diseases. Several tribal districts like Sidhi, Mandla, Shahdol, Khargone and others are rich in a variety of millets, maize, sorghum and cotton (including cultivation of unique coloured cotton). Several native breeds of fauna are also found in the area, including native breeds of cattle, buffalo and goat. There are 180 species of fish, out of which 40 are under the risk category, present mainly in the Narmada, Tapti and Mahi river basin area: Golden mahseer, Black rohu (carp) and catfish.

▲ ABOVE

The map depicts the protected area network of Madhya Pradesh.

CLIMATE

The state lies in the tropical zone. There are three distinct seasons in Madhya Pradesh – summer (March to May), monsoon (June to September) and winter (November to February). Both summer and winter are characterised by extreme temperatures. Madhya Pradesh receives the maximum rainfall from June to September. The eastern part of the state receives more rain (average 112 cm) as compared to northern and western areas (average 55 cm).

BIOGEOGRAPHIC ZONES

Two major mountain ranges of the state – Satpura and Vindhya – influence its geology and soil. Madhya Pradesh can be classified into two important biogeographic zones.

The north-western portion is semi-arid comprising of the Malwa Plateau. This consists of the Chambal drainage area in the north, Narmada watershed in the south, and the Betwa River in the east. Almost 98,700 sq km of the state falls in this zone, which amounts to 32 per cent of its area. Several national parks and sanctuaries are also part of this zone: Madhav National Park, Kuno-Palpur Sanctuary, National Chambal Sanctuary, Ghatigaon Wildlife Sanctuary, Karera Wildlife Sanctuary, Narsinghgarh, Gandhi Sagar, Sailana and Rala Mandal wildlife sanctuaries.

The Central highlands comprise of Satpura-Maikal and Vindhya-Baghelkhand ranges. Around 2,09,552 sq km of the state falls under this zone, which amounts to almost 68 per cent of its geographical area. This zone encompasses the well known tiger reserves: Kanha, Bandhavgarh, Pench, Satpura, Panna and Sanjay-Dubri.

The other protected areas of the zone include: Fossil and Van Vihar national parks, and several sanctuaries like Ratapani, Singhori, Son Gharial, Orchha, Gangau and Veerangana Durgavati. The protected areas in this zone occupy an area of 8,134 sq km.

AGRO-CLIMATIC ZONES

The state can be divided into the following agro-climatic zones:

- Satpura Hills and Kaymore Plateau
- Vindhyan Plateau
- Narmada Valley
- Wainganga Valley
- Gird (Gwalior) Region
- Bundelkhand Region
- Satpura Plateau
- Malwa Plateau
- Nimard Plateau
- Jhabua Hills

PROTECTED AREA NETWORK

The state has taken many initiatives for in-situ wildlife conservation. The protected area network comprises of nine national parks and twenty-five wildlife sanctuaries, covering an area of 10,862 sq km. This also includes six tiger reserves. The protected areas constitute almost 11.4 per cent of the state's total forest area and 3.5 per cent of its geographical area, which is still short of the national target (5 per cent of the total geographical area).

FORESTS

Madhya Pradesh is gifted with forests owing to its unique geographical and climatic attributes. All the forests are state owned. Various forest types ranging from dry thorn forests to tropical moist forests are found in the state. Pachmarhi has subtropical evergreen forests in higher elevations.

Almost one-third of the state's geographical area comprises of forests, which amounts to 95,222 sq km. However, dense forests occupy only 42,065 sq km of the total forest area.

The eastern part of the state has more forest cover than the northern and western parts. Most prominent tree species are teak, sal, bija, tinsa, saja and sesham. Bamboo is found as an understory in the dry and moist deciduous forests, on hilly slopes, valleys and stream banks.

The two most important tree species are found in the following districts:

Teak:
Jabalpur, Mandla, Chindwara, Seoni, Hoshangabad, Betul, Sagar, Panna, Dhar, Jhabua, Indore, Guna, Dewas and Sehore.

Sal:
Mandla, Balaghat, Sidhi, Shahdol and Jabalpur.

Mixed deciduous forests are found in Balaghat, Mandla, Chindwara, Hoshangabad, Nimar, Shivpuri, Shahdol, Sidhi and Rewa. These forests contain a large number of non-timber produce and medicinal plants. The important ones which are used by tribals and have commercial value as well are tendu leaves, sal seeds, mahua flowers, harra and gum.

EARLY DESCRIPTIONS ABOUT MADHYA PRADESH FORESTS AND WILDLIFE

The *Gazetteer of the Central Provinces of India* (1868) gives an account of the forests and wildlife in several districts of Madhya Pradesh covering Betul, Chindwara, Damoh, Hoshangabad, Jabalpur, Mandla, Nimar, Narsingpur, Sagar and Seoni. While describing Chindwara, it has been recorded,

> "The animals which are destructive to human life are the tiger, panther and bear, occasionally the hyena; there are in addition the hunting cheta, the wild dog and the wolf, which are destructive to flocks and herds; the wild boar, deer of all kinds, the samber, neilghay, and chetul, cause incessant damage to the crops. There are other wild animals such as foxes, jackals, and lynxes, & c. which keep down so successfully the quantity of small game in the district, that it is disproportionately scarce.

RIGHT

Kanha Tiger Reserve attracts hundreds of tourists annually, all of whom eagerly go on safari in hope of catching a glimpse of a tiger.

6

> "But there are hares, partridges and quail; and in the cold season the district is visited by the migrated birds, such as snipe, wild fowl, and the koolung, which later disappear after the gathering of the rubbee harvest. The bustard and florican are to be met with occasionally, but in no great numbers. In the Kumarpanee jungles the bison is to be seen, and also in the hills forming part of the Sautpoora range."

Several monumental books exist on the forests of Madhya Pradesh. Captain James Forsyth of the Bengal Staff Corps in his famous travelogue and sporting classic *Highlands of Central India* (1871), has described the forests and wild animals of Madhya Pradesh, the erstwhile Central Provinces.

Madhya Pradesh forms the meeting ground for many plant and animal life forms which are characteristic of north-eastern and south-western India. Sal, which mainly grows in the upper reaches of India is also found in the Satpura highlands of eastern Madhya Pradesh, whereas teak, a southern species, can be seen scantily in the eastern parts of the state.

Some interesting facts about the distribution of faunal species of north-eastern India have also been reported by Captain Forsyth. Along with sal, wild animals of the North East like the Wild buffalo, Barasingha or Swamp deer and the Red jungle fowl are also found in Central India (now Wild buffalo is a common sight in Chhattisgarh). However, these species are rarely seen beyond the sal belt in the western parts of the state.

▲ ABOVE

The silhouettes of bats flying is one of the most common sights at dusk in national parks across the state.

Captain Forsyth further records,

> "Two other large representatives of the eastern and western faunas, the wild elephant and the Asiatic lion, also appear to have formerly extended far into this region. In modern times, however, the advance of cultivation and the persecutions of the hunter have driven them both almost out of the country I am describing ..."

The higher hills comprise of subtropical forests, the western and southern areas have teak plantation, while sal occupies the moist eastern portions of the state.

WILDLIFE

The forests of Madhya Pradesh harbour a number of endangered faunal species: tiger, Barasingha, Chinkara, Black buck, Striped hyena, Lesser florican, Great Indian bustard, Wild dog. Some of these like bustard, florican, Black buck, Chinkara, are found in non-forested wildlife areas. The rivers of the state are home to Gharial and Mugger.

INDIGENOUS PEOPLE AND WILDLIFE

No account of Madhya Pradesh forest wealth is complete without appreciating the forest dwellers.

Madhya Pradesh has a rich tribal culture. Amongst the several tribes, Gonds and Baigas are prominent forest dwellers, apart from Pardhis. Gonds are found in the broken hills and forests of Betul, Chhindwara, Seoni, and Mandla districts. Many customs and food habits of these people are intimately linked with forests.

Baigas dwell in the Mandla and Balaghat districts of the state. Legends, folklores and beliefs of these people are intricately woven with the forests and wild denizens. When a Baiga is killed by a tiger, the spirit of the dead person is laid to rest through a special ritual.

The tribals believe this to be essential, as it not only prevents the ghost of the victim from residing on the tiger's head but also stops the animal from further human killing. The Baigas also believe that they can shut the jaws of a tiger to prevent it from harming them by driving a nail into a tree, which is done by a Baiga priest. Numerous such folklores abound in the areas of Kanha Tiger Reserve.

The Pardhis (Bahelia, Moghia and Shikari) are wandering fowlers and hunters. Their population spreads from Katni, Jabalpur and Panna to parts of Sheopur and Gwalior. The Pardhis are traditional hunters, who hunt all kinds of birds and smaller animals with snares (*phanda*). Different types of nooses are used to capture deer and Wild pigs. It has also been recorded that Pardhis excelled in using cheetahs to hunt Black buck in several states. Pardhis were deployed by village cultivators for crop protection from wild herbivores. The indigenous people also used the parts and derivatives of wild animals as drugs. Antlers of Barasingha were ground in water and applied on the chest and ribs for curing pain. The fat of Uromastix and Varanus was used for rheumatism. The flesh of Wild pigeon was used for paralysis.

TIGER RESERVES OF THE STATE

In the all-India tiger estimation using the refined methodology (2006), tiger presence has been reported in 15,614 sq km of the state and distributed in four regions – Pench (718 sq km), Satpura (12,700 sq km), Bandhavgarh (2,000 sq km) and Kanha (3,162 sq km). The tiger population in other parts of the state is very small. Low density tiger presence has been found all along the forests in the north of Narmada from Jabalpur to West Nimar. Their habitats constitute crucial linkages which need to be conserved by ensuring ecologically sustainable land usage, through landscape level land use planning.

This necessitates mainstreaming tiger conservation in other areas, to integrate the same in various production sectors, where the primary emphasis is not wildlife or tiger conservation.

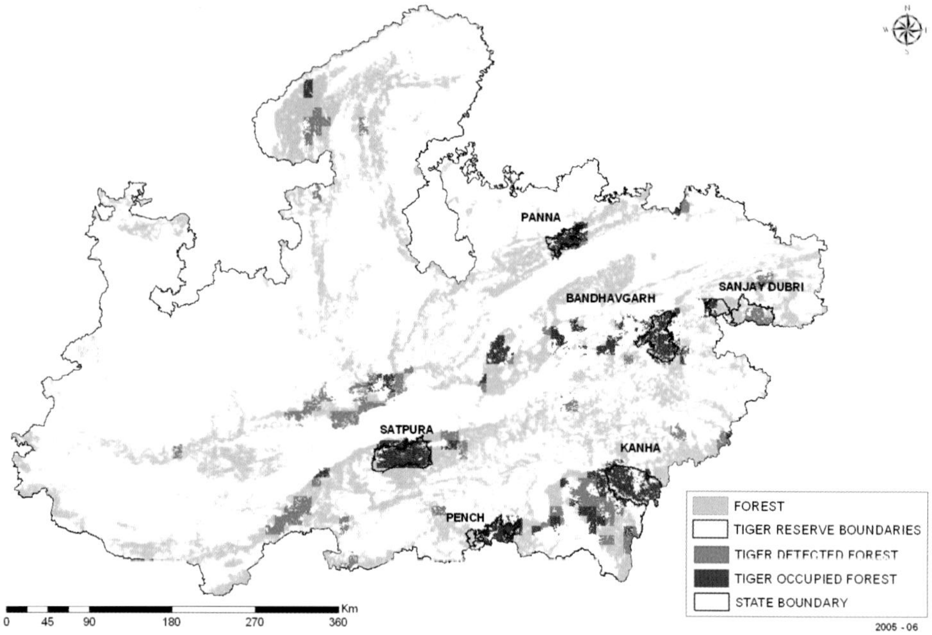

CONSERVATION INITIATIVES IN MADHYA PRADESH

Madhya Pradesh has the distinction of having one of the first nine tiger reserves (Kanha) in the country, when Project Tiger was initially launched in the early seventies.

At present, there are six tiger reserves in the state. Wildlife conservation in India is a shared responsibility between the Central government and states. The tiger reserves of Madhya Pradesh receive considerable funding support from the ongoing Centrally sponsored schemes like Project Tiger and Integrated Development of Wildlife Habitats, apart from investments made by the state.

The strategic conservation initiatives taken by the state include creating an effective protected area network of 10,862 sq km which covers nine national parks in twenty-five wildlife sanctuaries (six of which are designated as core/critical areas of tiger reserves).

Several externally aided projects like the M.P. Forestry Project, and the India Eco-development Project have also resulted in strengthening the infrastructure and capacity building of protected areas, apart from fostering inclusive agenda involving local people. The state has taken special initiatives to elicit local public support by fostering village level micro institutions like Forest Protection Committees and Eco-development Committees.

▲ ABOVE

White pelicans taking flight from one of the lakes in Kanha Tiger Reserve.

▲ ABOVE LEFT

The map depicts the distribution of tigers across different tiger reserves in Madhya Pradesh.

The crucial steps taken in the state for wildlife conservation include curbing poaching and illegal wildlife trade, habitat improvement, structural support, eco-development, estimation of tigers and other wild animals using refined methodology, management planning for protected areas, research and wildlife health care with field units, protection of wild animals outside the protected areas. The state government has set up a Tiger Foundation Society and Madhya Pradesh Eco-tourism Development Board. Schemes for conserving Lesser florican, Great Indian bustard have been launched and earnest initiatives to conserve Mugger and Gharial have been undertaken by the state authorities.

Madhya Pradesh has the unique distinction of being the first state to create a Tiger Cell at the district and state level for strengthening tiger protection. It is also the first state to recycle the gate receipts from protected areas for creating a fund (Vikas Nidhi) to benefit the protected areas staff and local people. The recent initiative taken by the state in creating a Tiger Strike Force with a headquarter and five regional units is also noteworthy.

The state has also developed the Kuno-Palpur Sanctuary, in collaboration with the Government of India, as an alternate home for rearing the endangered Asiatic lion. As an outcome of concerted efforts under Project Tiger, the Central Indian Barasingha has been saved from the brink of extinction and is now thriving in the Banjar and Halon valleys of Kanha Tiger Reserve. The wildlife corridor between Kanha Tiger Reserve and Achanakmar Tiger Reserve of Chhattisgarh has been identified and evaluated by the Wildlife Institute of India, Dehradun in the G.I.S. (Geographical Information System) domain. Efforts are underway for strengthening the corridor between Kanha and Pench.

WILDLIFE CONSERVATION : A COLLECTIVE RESPONSIBILITY

Wildlife conservation is a collective responsibility between state governments, Government of India and civil society.

In our federal set up, the Government of India provides enabling statutory provisions, technical guidance and funding support for in-situ and ex-situ conservation of wildlife in states. The day-to-day management of protected areas and tiger reserves and deployment of staff rest with states, who own the forests.

Several milestone initiatives have been taken by the Government of India for wildlife conservation. These include, implementing the urgent recommendations of the Tiger Task Force, creating the National Tiger Conservation Authority, establishing the Wildlife Crime Control Bureau and revising the ongoing Centrally sponsored schemes to support states in conservation initiatives by

providing funding support. Ongoing schemes like Project Tiger, Project Elephant, Integrated Development of Wildlife Habitats are supporting states including Madhya Pradesh in a big way for in-situ conservation of wild animals.

An area of around 30,000 sq km of core/critical tiger habitat has been identified in states as inviolate to foster tiger conservation after village relocation. The relocation package for people has also been enhanced to Rs 10 lakhs per family (from the earlier Rs 1 lakh per family) thereby ensuring a fair deal to such people. Almost 47 per cent of Project Tiger allocation goes for supporting anti-poaching operations and strengthening protection infrastructure.

The recent all-India tiger estimation using the refined methodology in the G.I.S. domain has revealed that the source population of tigers in tiger reserves is still secure in the country including Madhya Pradesh, owing to the efforts taken under Project Tiger. However, these areas need active managerial interventions. The status of tigers and other wild animals outside the tiger reserves and some protected areas in the country is not viable. This can be attributed to several causal factors: mortality of wild animals due to poaching and man-animal conflicts, degradation of the forest status outside protected areas/tiger reserves owing to human pressure, livestock pressure and ecologically unsustainable land uses and fragmentation leading to a loss of gene flow from source populations. Adding to these are problems like the loss

▲ ABOVE

The Common mongoose is a carnivorous mammal that is easily spotted in almost all national parks of Madhya Pradesh.

▶ RIGHT

A herd of Chital, or Spotted deer, grazing in Kanha Tiger Reserve. Spotted deer always move in crowds. By making frequent alarm calls, they are always on the alert for predators.

of reproduction owing to disturbances on account of heavily used infrastructure like highways, a lack of adequate protection to the animals, loss of forest quality in terms of prey biomass to support large carnivores like tigers and leopards. Insurgency or law and order problems in some tiger reserves/ protected areas/forest areas also pose danger to the wildlife.

Stepping up protection in tiger reserves, creating buffer zones around core/critical tiger habitats, promoting low profile, regulated eco-tourism in protected areas, reviving corridor connectivity outside the source population areas, mainstreaming wildlife concerns in sectors where conservation is not the primary goal and ensuring ecologically sustainable land uses in identified wildlife corridors are some of the challenges the state faces. With the active participation of the Centre, state and civil society institutions these challenges can be easily met.

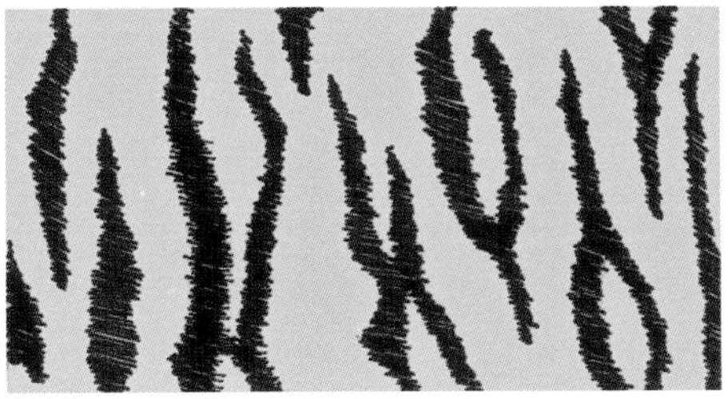

TIGER RESERVES

- BANDHAVGARH
- KANHA
- PANNA
- PENCH
- SANJAY-DUBRI
- SATPURA

LEFT

A family enjoys an elephant safari — probably the best way to experience a national park — in Panna Tiger Reserve.

BANDHAVGARH

Bandhavgarh is located between the Vindhya and Satpura ranges, and was well known for its tigers much before it became a tiger reserve. The area has unique historical attributes, with the Bandhavgarh Fort prominently located as a major landmark. According to legend, it is believed that the Bandhavgarh Fort was given to Lakshman by his brother Lord Ram (*bandhav* means brother, hence the name Bandhavgarh). Several ancient texts like *Shiva Samhita* and *Narad Puran* also make a reference to this area.

Many dynasties ruled Bandhavgarh: the Vakatakas (third century onwards), the Sengars (fifth century onwards) and the Karchulis (twelveth century onwards). During the thirteenth century the Baghela Rajputs took over the fort, which remained their capital till 1617. Subsequently, Maharaja Vikramaditya Singh moved his capital to Rewa, and the fort was deserted in 1935. Several descriptions, relief images, rock cut caves and statues adorn the Bandhavgarh Fort plateau and its surrounding areas.

◀ LEFT
A visitor enjoys the expanse of Bandhavgarh's natural beauty. Binoculars are an essential part of tiger spotting and bird watching.

Bandhavgarh owes its wilderness to erstwhile rulers of the Rewa State, since it was their hunting preserve and enjoyed protection from wanton destruction. It was considered a good omen for the Rewa Maharaja to shoot 108 tigers, the numbers coinciding with the beads in the Hindu rosary!

It is said that the poet saint Kabir stayed in the Bandhavgarh Fort during the reign of Maharaja Vir Bhanu Singh. Later, Maharaja Ram Chandra of

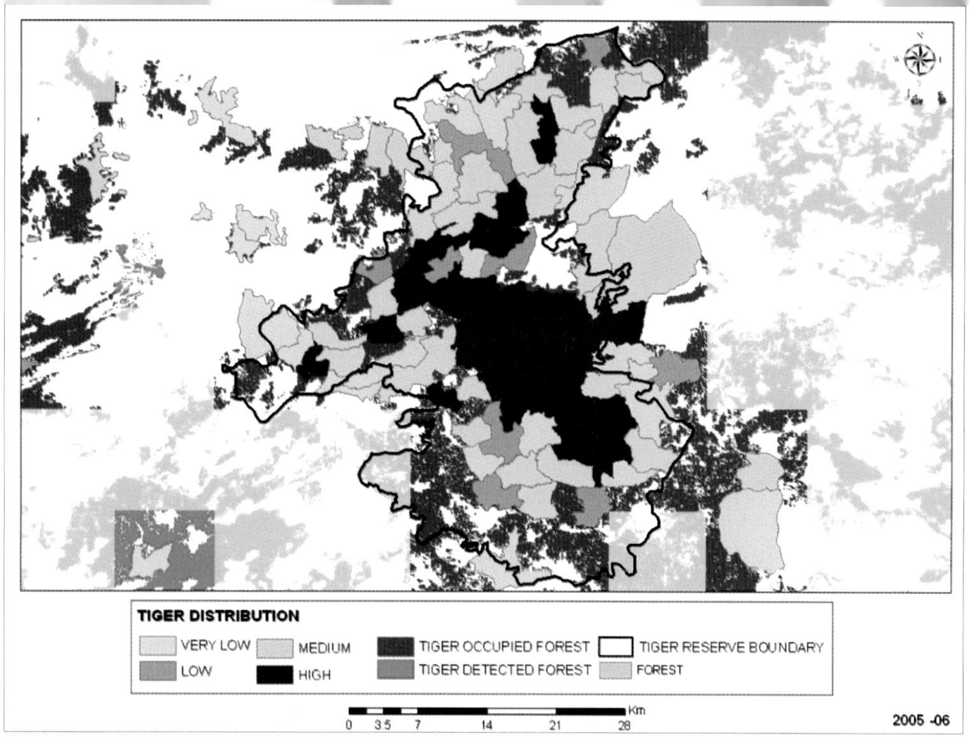

Rewa dynasty, gave protection to the Begum of Emperor Humayun during the reign of Emperor Sher Shah Suri. It is also said that the Maharaja introduced his musician Tansen to the court of Emperor Akbar. The Rewa State merged with Vindhya Pradesh in 1947 and new regulations were enforced. During 1959 the Madhya Pradesh Game Act came into force, and an area of 105 sq km (Tala range) was notified under the said Act as a national park in 1968.

Bandhavgarh thus has three major attractions – wildlife, archaeology and breathtaking landscape.

The Bandhavgarh-Rewa-Sidhi landscape of Rewa State was also home to White tigers. The last known White tiger, Mohan, was caught by Maharaja Martand Singh in 1951, some 60 km away from Bandhavgarh.

Bandhavgarh is characterised by valleys, hills and plains. Numerous hillocks, big and small, are found in the reserve with Bandhavgarh and Badhaini prominently towering amongst them. The area also forms part of the catchment of several rivers like Charanganga, Janad, Umrar and Chachahi.

The forests are of tropical moist deciduous type, consisting of sal and mixed forests with grasslands. Sal is found in the lower hill slopes and valleys. Mixed deciduous forests are found in the hills and drier parts of the reserve. Bamboo is

▲ ABOVE

Bison or Gaur was once commonly seen in Bandhavgarh during the months of March and April.

▶ RIGHT

The Brown shrike is only one of the migratory birds easily spotted in Bandhavgarh.

a common understory. The grassy patches are locally known as *vahs*, which occur along perennial streams. Raj Behra is a prominent, marshy grassland of the reserve.

The tiger is the star attraction, and can be easily sighted in its various moods and behavioural patterns. Interesting territorial fights and land tenure patterns of tigers have been observed in the area. The other common fauna include the leopard, Wild dog, and common wild herbivores of Central India like Sambar, Chital (Spotted deer), Chowsingha, Nilgai and Chinkara. Other wild animals include the Sloth bear, jackal, hyena, langur, and more than 240 species of avifauna, apart from a large number of reptilian species. The reserve also had a small population of Gaur till the mid-nineties, considered the only population thriving in the north of Narmada River. Although, very rarely seen, recently, there have been reports of Gaur being sighted in forest areas adjoining the reserve.

The must-visit places within the park include Bandhavgarh Fort, the statue of reclining Vishnu (*Sheshshaiya*), Badi Gufa, Three Cave Point, Sita Mandap, Raj Behra and Chakradhara. Wildlife can be viewed from elephant back or open vehicles.

FOR VISITORS

Fauna:
Tiger, panther, Wild dog, Sambar, Chital, Sloth bear, Wild boar, langur along with quite a few species of mammals and reptiles, besides species of avifauna.

Flora:
Sal, saja, mahua, achar, amla, palas, arjun, dhaora, lendia and bamboo, apart from many species of shrubs, herbs and grasses.

Archeological attractions:
The natural beauty is supplemented with remnants of an ancient past including caves, rock paintings, carvings and a fort.

Best Season:
February to June.

BANDHAVGARH
TRAVEL GUIDE

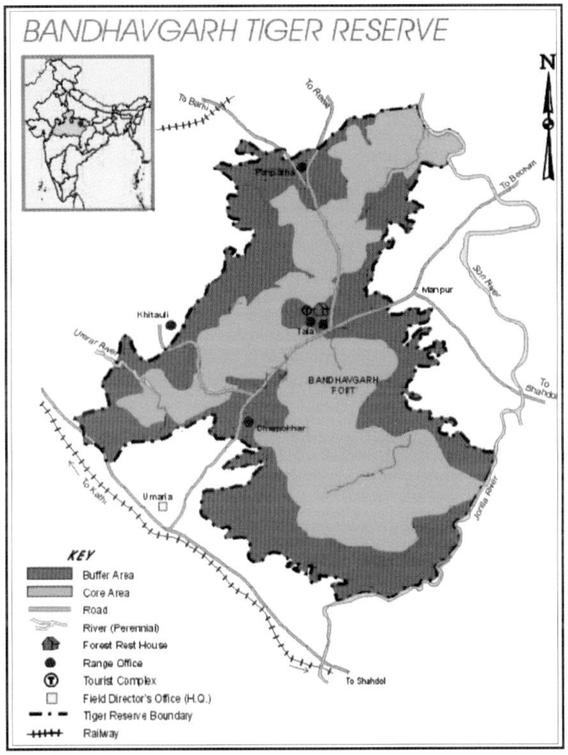

LEFT

The eleventh-century statues of various incarnations of Lord Vishnu, carved out of monolith rocks, are found in the Bandhavgarh Fort. Sheshshaiya, *the statue of Lord Vishnu in reclining pose on a bed of snakes, is the largest of them all.*

Contact

Field Director
Bandhavgarh Tiger Reserve,
Umaria - 484661, Madhya Pradesh
Telephone No. (07653) 222214 (O)
E-mail: fdbtr@rediffmail.com

Core Area	: 716.903 sq km
Longitude	: 80° 47'15" to 81° 11'45" E
Latitude	: 23° 30'12" to 23° 45'45" N
Altitude	: 440 m to 810 m above msl
Rainfall	: 1,175mm

Temperature

Minimum	: 5° C
Maximum	: 44° C

Seasons

Summer	: Mid-April to mid-June
Monsoon	: Mid-June to September
Winter	: November to mid-February

The reserve is closed from the 1st of July to 15th of October for visitors every year (dates are subject to change).

Reaching there

By rail

The nearest railway stations are Jabalpur (190 km), Satna (130 km) and Katni (102 km) in the Central Railway zone and Umaria (35 km) in the South Eastern Railway zone.

By air

The convenient airports for reaching Bandhavgarh are Jabalpur or Khajuraho, from where the journey by road takes four and five hours respectively.

Accommodation

Available at the White Tiger Forest Lodge (MPSTDC), Forest Department and private lodges.

KANHA

Kanha Tiger Reserve is located in the Maikal Hills of the Satpuras spreading over two revenue districts – Mandla and Balaghat. Kanha has a long history of conservation. It was declared a reserve forest in 1879 and notified as a wildlife sanctuary in 1933. Its status was further upgraded to a national park in 1955. The habitat has an excellent interspersion of geographical attributes and welfare factors which foster a rich population of wildlife.

The flat hill tops or plateaus are known as *dadars*, which are juxtaposed with grassy meadows, sal mixed forests and riverine patches. The *dadars* have scanty tree growth like aonla, achar and tendu with plenty of grass growth. The vegetation in valleys is moist with large mango trees having bee hives, interspersed with bamboo growth.

◂ LEFT

During the hot summer months, the only respite for tigers is to cool off in water bodies dotted around Kanha Tiger Reserve.

Maikal range is the prominent hill feature, which forms the watershed between the rivers Narmada and Mahanadi. The hill range in the western portion of the reserve is known as the Bhaisanghat Ridge, which divides the catchment area of Narmada between the eastern (Halon) and western (Banjar) parts of the reserve. Banjar and Halon are the main rivers flowing through the reserve. The central meadows of Kanha are encircled in a horseshoe shape by hill formations, leaving a northern gap, which is the Sonf Meadow. The discontinuous distribution of gregarious sal patches intermixed with grassy banks provide ideal edges for Spotted deer.

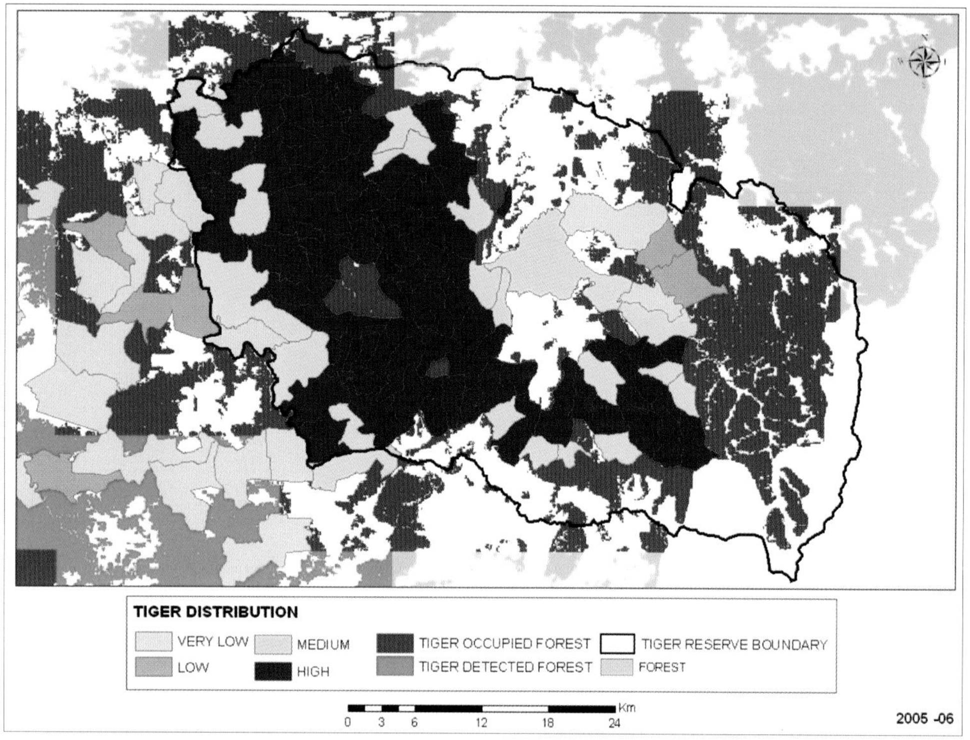

Kanha has a relocation history. As many as twenty-seven forest villages have been relocated from the core area of the reserve under Project Tiger. These relocated sites, along with abandoned shifting cultivation areas of the past, and the ground frost areas provide heterogeneous grasslands (meadows) for wild animals. Managerial interventions aim to arrest such meadows in the grassland stage to foster wild ungulates.

Kanha is virtually a tiger land, with several tigresses occupying traditionally famous natal areas. Several pockets of high density areas lead to competition amongst tigers for food and space with several signs indicating their presence. Inter group fights amongst tigers and cub mortality caused by male tigers are common in Kanha.

There is a rich assemblage of co-predators and prey animals. Packs of Wild dogs chasing large herds of Spotted deer are a common sight in Kanha. Perhaps the most precious animal of Kanha is the Central Indian Barasingha. This Swamp deer is the last world population of the hard ground subspecies, which has virtually been saved from extinction owing to concerted efforts under Project Tiger. Barasinghas have also been located successfully to the eastern Halon Valley of the park, which was their original home.

▲ ABOVE

Young Spotted owlets huddle for warmth.

▸ RIGHT

An elephant safari.

Other common animals found in the reserve are typical of Central India, which include the leopard, Spotted deer, Sambar, Wild pigs, Grey langur and more than 300 species of birds.

Kanha is the pride of Project Tiger with ongoing managerial interventions having considerable demonstrative value for others to emulate. The creation of an inviolate core area through successful village relocation, developing a functional buffer zone under the unified control of tiger reserve management, setting up of seasonal protection strategies, habitat management, and interpretive facility are notable achievements of this reserve.

The places of tourist attraction are Shravan Tal, Shravan Chita, Macha Dongar, Bamni Dadar famous for breathtaking sunset view, orientation centre/museum and grasslands of Mukki and Sonf.

Wildlife viewing from elephant back is a major attraction for visitors, apart from jungle excursions in open vehicles.

FOR VISITORS

Fauna:
Tiger, panther, Wild dog, Gaur, Barasingha, Sambar, Chital, Chowsingha, Nilgai, Sloth bear, Wild boar, langur and many other species of mammals and reptiles, besides species of avifauna.

Flora:
Sal, saja, bija, jamun, mahua, semal, amla, tendu, dhaman, palas, kusum, arjun, dhaora, harra, bahera, lendia and bamboo, apart from numerous grass species.

Best Season:
February to June.

KANHA
TRAVEL GUIDE

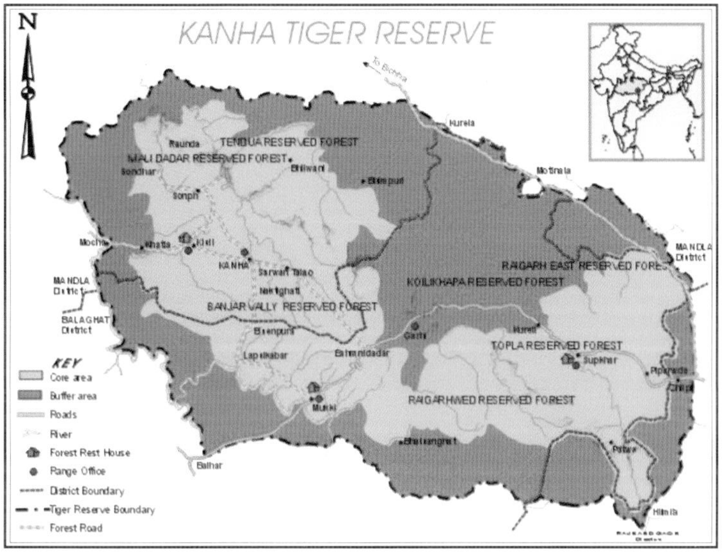

Contact
Field Director
Kanha Tiger Reserve
Madhya Pradesh - 481661
Telephone No. (07642) 250760 (O)
Fax No. (07645) 250761
E-mail address: fdkanha@rediffmail.com

LEFT
Clusters of sal forests are the most common flora of Kanha.

Core Area	: 917.43 sq km
Longitude	:80 ° - 26'-10" to 81 ° - 4'- 40" E
Latitude	: 22 ° -1'- 5" to 22 ° - 27'- 48" N
Altitude	: 450 m to 950 m above msl
Rainfall	: 1,224 mm

Temperature

Minimum	: 2 ° C
Maximum	: 43 ° C

Seasons

Summer	: March to mid-June
Monsoon	: Mid-June to early October
Winter	: November to February

The reserve is closed from the 1st of July to 15th of October for visitors every year (dates are subject to change).

Reaching there
By air
Nagpur (266 km) and Jabalpur (175 km) are the convenient airports.

By rail and road
Jabalpur (175 km) is the convenient railhead to visit Kanha. By road Kanha National Park is well connected with Jabalpur (175 km), Khajuraho (445 km), Nagpur (266 km), Mukki (25 km) and Raipur (219 km).

Accommodation
Available at Baghira Log Huts and Tourist Hostel, Kisli and Kanha Safari Lodge, Mukki (MPSTDC), Forest Department and private lodges.

PANNA

Located in the Vindhya range, this tiger reserve spreads into two revenue districts, Panna and Chhatarpur. Panna links the eastern and western wildlife habitats in the state through the Vindhya range, running from the north-eastern side to the south-western portions.

Like Bandhavgarh, Panna was also the hunting reserve of its erstwhile rulers of Chattarpur and Bijawar princely states. The topography of Panna is unique, comprising of three distinct tablelands or plateaus – the upper Talgaon Plateau, the middle Hinouta Plateau and the Ken Valley with undulating hills.

Numerous gorges, waterfalls and valleys characterise the Panna landscape. The Ken River flows through the reserve from north to south, and is its lifeline. Several small rivers also join this river.

LEFT

The grasslands of Panna Tiger Reserve serve a dual purpose – they provide food to herbivores and help the carnivores camouflage themselves while hunting.

The Gangau Wildlife Sanctuary comprising of forests of Panna Forest Divisions (north and south) was created in 1975. The area was enlarged and its legal status elevated to a national park during 1981. The protected area was brought under Project Tiger in 1994.

The vegetation consists of tropical dry teak and dry deciduous mixed forests. The dry forests are mainly found in sandstones and shallow soils. Many gregarious patches of salai trees are commonly seen in the area.

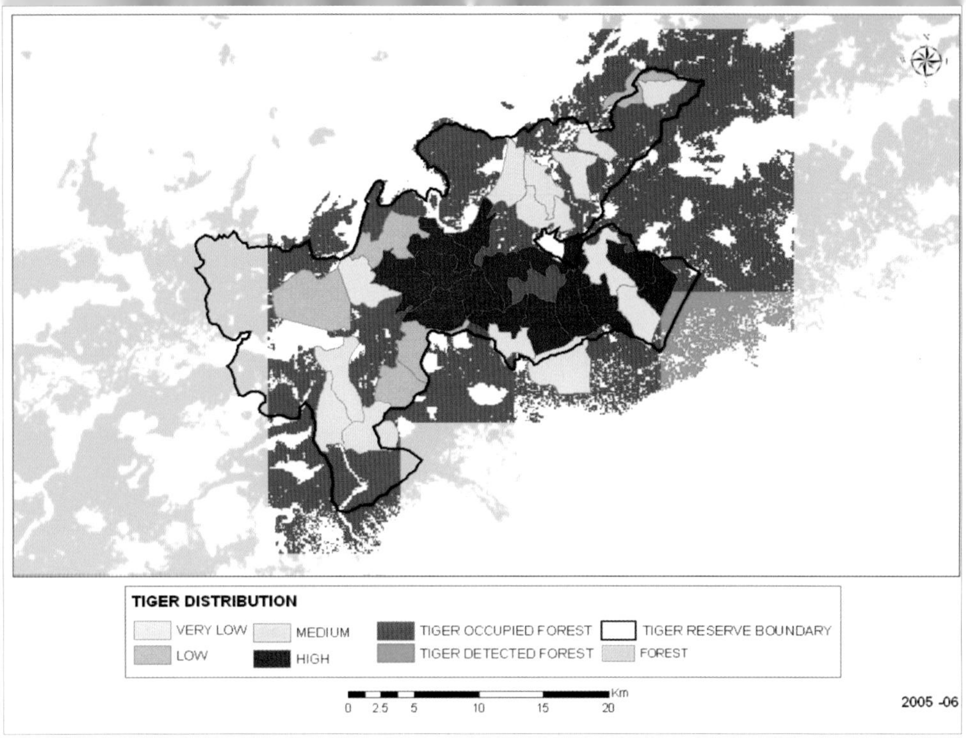

The reserve forms the eastern limit of teak-kardhai mixed forests. It is also the northern limit of the natural distribution of teak.

Panna is rather an open, dry landscape, with the topography and water distribution governing the spatial presence of wild animals. There is a significant drop in elevations, locally known as *seha*. These places provide adequate cover for wild animals.

Vultures and other raptors roosting on cliffs along the river course are a common sight in Panna. The fauna consists of tiger, leopard, Chowsingha, Sloth bear, Nilgai, Chinkara, hyena and small groups of Chital. More than 200 species of birds have been spotted. Both Gharial and Mugger are found in the Ken River.

The management focuses on protection, water augmentation and control of biotic pressure through village relocation, with viable livelihood options to local people.

The places of tourist interest are Pandava Fall, Kamasan Fall, Dhundhawa Fall, Bahuradeh Fall, Bhadar and Badgadi Fall, Raneh Fall and cave paintings.

ABOVE RIGHT

Entrance to Panna Tiger Reserve.

RIGHT

Asian pied starlings (Pied mynah) in Panna.

FOR
VISITORS

Fauna:
Tiger, panther, wolf, Wild dog, hyena, Sambar, Chital, Chowsingha, Chinkara, Nilgai, civet, Rhesus monkey and porcupine along with many species of snakes and fishes; apart from species of birds including the Paradise flycatcher, the state bird of Madhya Pradesh.

Flora:
Teak, khair, kardhai, tendu, dhawa and bamboo along with many species of grasses.

Best Season:
December to May

PANNA
TRAVEL GUIDE

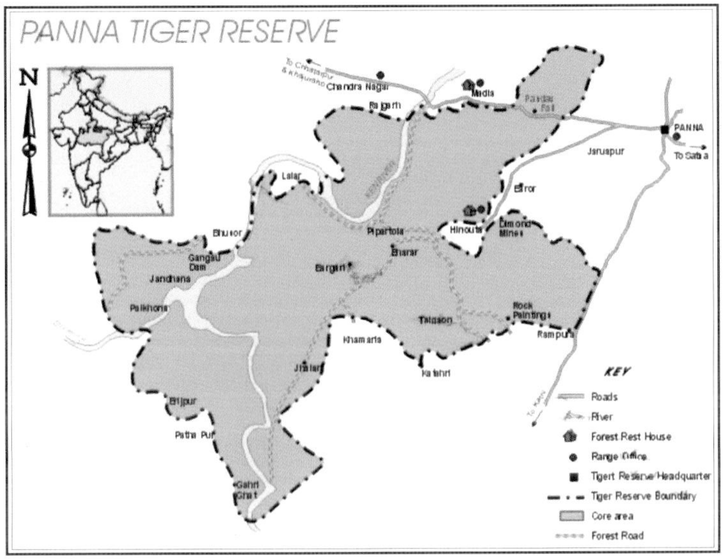

Contact
Field Director
Panna Tiger Reserve
Panna
Madhya Pradesh- 488001
Telefax No. (07732) 291214 (O)
E-mail address : pannatr@sancharnet.in

LEFT
The fawn coloured coat of the Chital works well to camouflage it in its surroundings.

Core Area : 576.13 sq km
Longitude : 79° 45' to 80° 09' E
Latitude : 24° 27' to 24° 46' N
Altitude : 330 m above msl
Rainfall : 1,100 mm average

Temperature
Minimum : 5° C
Maximum : 45° C

Seasons
Summer : March to mid-June
Monsoon : Mid-June to mid-September
Winter : Mid-November to February

The reserve is closed from the 1st of July to 15th of October for visitors every year (dates are subject to change).

Reaching there
By rail
Satna and Jhansi at a distance of 90 km and 200 km respectively are nearest railway stations.

By air
The nearest airport is at Khajuraho (25 km).

By road
The park is well connected with other parts of the region by a good network of roads.

Accommodation
Available with MPSTDC and private lodges.

PENCH

Named after the Pench River, which flows through the reserve, this tiger habitat is located in the southern reaches of the Satpura hill range, falling in the Seoni and Chhindwara districts. The terrain has small hill ranges with a dense forest cover.

The area was declared a sanctuary in 1977 and portions of it were elevated as a national park in 1983. During 1992 these areas were brought under the coverage of Project Tiger. The area has been described in several wildlife books, including the travelogue of Captain J. Forsyth, *Highlands of Central India*, R. A. Sterndale's *Seonee: Camp life on Satpura Hills*, and A.A. Dunbar Brander's *Wild Animals in Central India*.

◂ LEFT

The terrain of Pench Tiger Reserve is undulating, with most of the area covered by small hill ranges and dotted with water bodies.

The landscape is also the setting for Rudyard Kipling's famous *Jungle Book* and its Mowgli stories. It is believed that Kipling was inspired by Sir William Henry Sleeman's account of a boy believed to have been raised by wolves, in the village of Sant Baori in the Seoni district. Several locations described in the *Jungle Book* are found in the area – the Wainganga River with its gorge where Sherkhan was killed, 'Seeonee' and the Kanhiwara hills. Pench has an undulating terrain with gentle slopes and seasonal rivulets. The Pench River flows through the middle of the core area of the tiger reserve. During summer the river is dry, but a number of pools (locally known as *dohs*) provide water for wild animals.

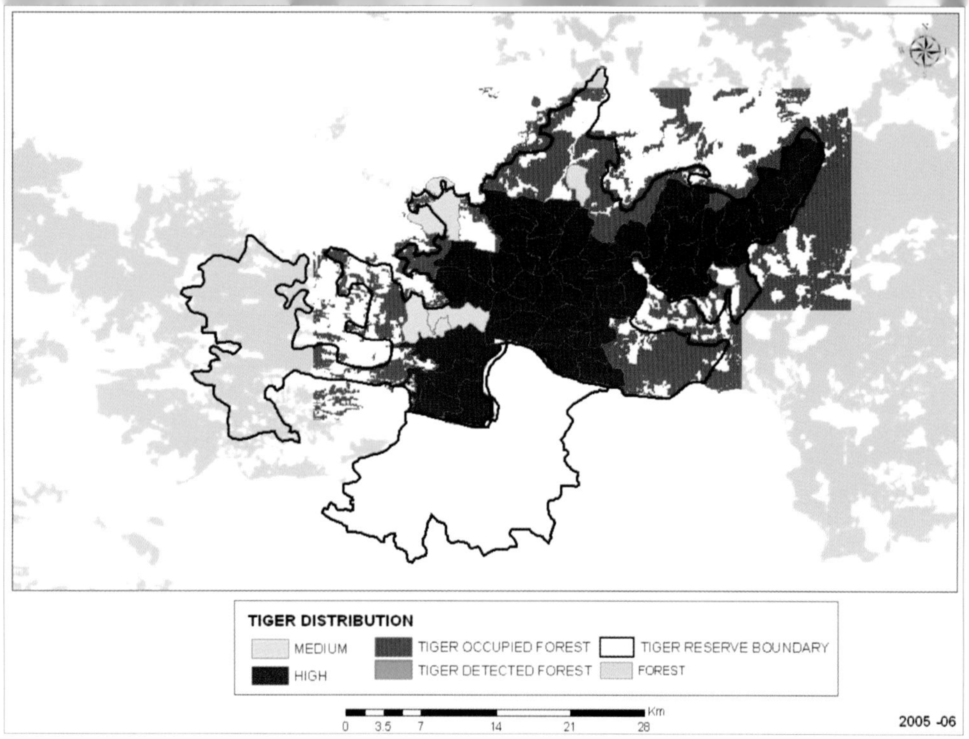

The Pench hydroelectric dam is located at the Maharashtra-Madhya Pradesh border, and a considerable portion of the submerged area (54 sq km) falls within the reserve. The forests are continuous with the adjoining Pench Tiger Reserve of Maharashtra. Pench-Kanha landscape is important for tiger conservation, since the landscape connects two major sources of population of tigers in the state. The vegetation comprises of moist and dry teak along with mixed deciduous forests. As many as 1,200 species of plants have been recorded from the area, which includes species of ethno botanical importance as well.

The reserve has a good tiger population, and the litter size more often is big enough to have four cubs. Co-predators like the Wild dog and leopard are very common. The herbivores include Sambar, Chital, Nilgai, Chowsingha, Chinkara, Barking deer, Gaur and Wild pig. The most commonly found omnivore is the Sloth bear. More than 280 bird species have been check listed. The common ones include cormorants, darters, herons, egrets, White neck storks, White ibises, Black ibises, Pintail ducks, Brahminy ducks, Fishing eagles, shrikes and wagtails. Four species of vultures – the White rumped, Long billed, White scavenger and the King are seen in the reserve.

The must-visit places include Totladoh Dam, Karmajhiri, Kalapahad, Chhindimatta, Bison Camp and Bison Retreat.

ABOVE RIGHT

Grey langurs spend most of their time on the ground, though they sleep on trees.

RIGHT

Bison or Gaur is primarily found in deciduous and evergreen forests, but their existence is threatened due to the depletion of the forest cover.

FOR
VISITORS

Fauna:
Tiger, panther, Wild dog, Gaur, Sambar, Chital, Nilgai, Sloth bear, Wild boar, langur and many other species of mammals and reptiles, besides species of avifauna.

Flora:
Teak, saja, bija, jamun, dhaora, mahua, semal, lendia, amla, tendu, dhaman, palas, kusum, arjun, bahera and bamboo, apart from many species of grasses.

Best Season:
December to April.

PENCH
TRAVEL GUIDE

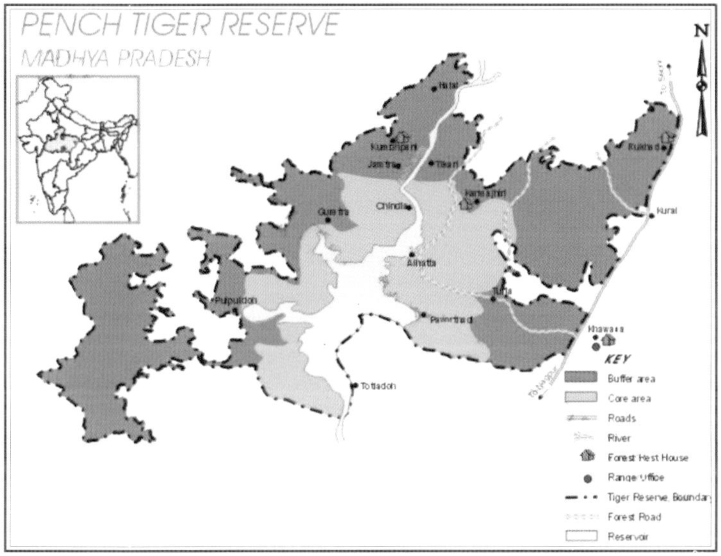

Contact
Field Director
Pench Tiger Reserve
District Seoni - 480661
Madhya Pradesh
Telephone No. (07692) 223794 (O)
E-mail: penchtr@sancharnet.in

◂ LEFT
A tiger rests in the shade.

Core Area	: 411.33 sq km
Longitude	: 79° 7'45" to 79° 22'30" E
Latitude	: 21° 37" to 21° 50'30" N
Altitude	: 580 - 675 m above msl
Rainfall	: 1,397 mm

Temperature

Minimum	: 4° C
Maximum	: 47° C

Seasons

Summer	: March to mid-June
Monsoon	: Mid-June to September
Winter	: November to February

The reserve is closed from the 1st of July to 15th of October for visitors every year (dates are subject to change).

Reaching there
Pench is only 12 km away from Khawasa on N.H. 7 between Nagpur and Jabalpur; Khawasa is 81 km from Nagpur, and 193 km from Jabalpur.

The convenient air and rail heads are Nagpur and Jabalpur.

Accommodation
Available at Kipling's court (MPSTDC), Forest Department and private lodges.

SANJAY-DUBRI

Sanjay-Dubri landscape is characterised by undulating hills, valleys, deep gorges and plains with numerous streams. The reserve is located in the Sidhi district, which was the habitat of Mohan, the last known White tiger living in the wild. Various perennial streams flow through the river – Gopad, Mahan, Kodmar, Banas, Umrari, Magdar, Bijour and Patnaiya. The areas around these abound in riparian vegetation.

The vegetation comprises of northern tropical sal and dry deciduous forests. Sal dominates the reserve with gregarious patches, apart from mixed forests and bamboo. As many as eighty-two species have been recorded in the reserve.

Sanjay is an important link in the north-eastern part of the state between Bandhavgarh and Palamau tiger reserves. The forests to the northeast of Bandhavgarh are continuous with the forests of Sanjay Tiger Reserve. However, several areas require restorative management to ensure corridor connectivity. The tiger reserve also forms the watershed of the Son River and its tributaries.

The fauna consists of tiger, leopard, hyena, jackal, Chital, Sambar, Chinkara, Sloth bear, Wild pig, langur and a large variety of birds. Sanjay also forms the seasonal migratory route for wild elephants from Chattisgarh.

ABOVE RIGHT

Grey langur perched on a tree.

RIGHT

The Asian elephant is the largest land mammal of the subcontinent.

In 2008, Sanjay Tiger Reserve was brought under Project Tiger. Securing inviolate core area for tiger and other wild animals by reducing the biotic disturbance, and stepping up field protection are the challenges before the reserve's management.

The must-visit places include Ramadaha Kund, Domarpert Hill and Gopad River.

FOR VISITORS

Fauna:
Tiger, leopard, Jungle cat, hyena, jackal, fox, Chital, Sambar, Chinkara, Barking deer, Wild boar, Sloth bear, langur and Indian pangolin.

Flora:
Sal, saja, dhaora, salai, bija, haldu, tendu, aonla, lendia, tinsa and bamboo.

Best Season:
December to April.

SANJAY-DUBRI
TRAVEL GUIDE

Core Area	: 831.250 sq km
Longitude	: 81°30" to 82°15" E
Latitude	: 23°46" to 24°15" N
Altitude	: 770 m above msl
Rainfall	: 1,010 mm

Temperature

Minimum	: 4° C
Maximum	: 47° C

Seasons

Summer	: March to mid-June
Monsoon	: Mid-June to September
Winter	: November to February

The reserve is closed from the 1st of July to 15th of October for visitors every year (dates are subject to change).

Reaching there
Varanasi (210 km) and Jabalpur (220 km) are convenient air/rail heads for reaching Sanjay Tiger Reserve.

Accommodation
Available only with Forest Department.

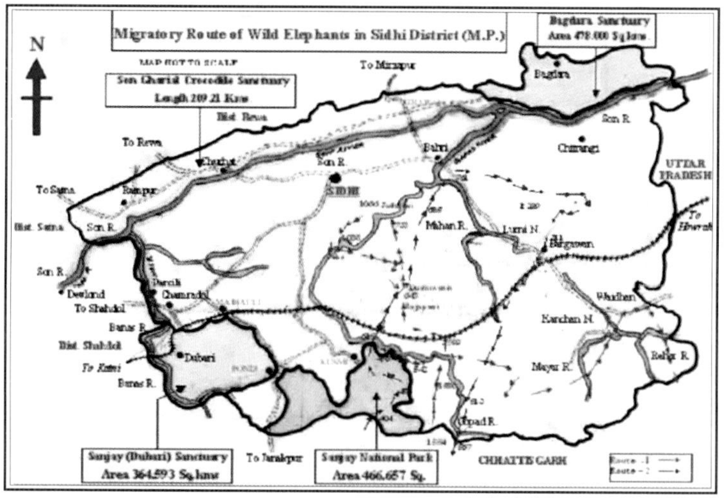

Contact
Field Director
Sanjay-Dubri Tiger Reserve
Sidhi - 486661
Madhya Pradesh
Telefax No. (07822) 250121 (O)
E-mail: snpsidhi@sancharnet.in

SATPURA

The Satpura Tiger Reserve comprises of large plateaus, with Dhoopgarh Peak towering to almost 1,352 m above sea level, and Pachmarhi at its highest point. Other elevated, well known hill tops include Chauragarh, Mahadev, Jambudweep and Bee Fall. Valleys, deep gorges, plains and water bodies characterise the Satpura landscape.

Satpura is a well known Gondwana tract. The Gond tribes predominantly occupied this area, practising shifting cultivation. During the year 1862, the Forest Department of Central Provinces constructed the Bison Lodge at Pachmarhi. Though the original construction no longer exists, a forest museum has been created at this spot.

Satpura is an important habitat, harbouring a source population of tiger, which disperses to the adjoining Betul, Hoshangabad and East Nimar areas.

The Tawa Reservoir containing the impounded water of Tawa River is situated on the north-western side of the tiger reserve, with its backwater spreading over to almost 200 sq km of the reserve area.

The forest vegetation is interesting. The tract is considered as a relict site, representing some of the floral species common to Western Ghats and the Himalayas. Perhaps, as considered in the 'Satpura hypothesis', the area in the distant geological time scale might have been a bridge, linking

portions of Western Ghats with the Himalayas. Some portions of the reserve have endemic flora, with a large number of tree and aquatic ferns, species like Berberis, apart from insectivorous plants like Drosera.

The forests, by and large, comprise of tropical moist deciduous and tropical dry deciduous vegetation, with large areas of teak and its associates, with sal trees in the plateaus.

The faunal assemblage is rich with tiger and its co-predators like leopard and Wild dog. The herbivores include Gaur, Sambar, Chital, Barking deer, Nilgai, and Chowsingha. Arboreal animals like Giant and Flying squirrels are also frequently seen along the riverine areas. Fresh water crocodiles and fishes are common in the water bodies.

Red jungle fowls, partridges, quails, parakeets and raptors are part of the rich avifauna of the area.

The places of tourist interest include Jata Shankar, Pandav Caves, Priyadarshini Point, Mahadev, Chauragarh, Dhoopgarh, Cave Shelter, Bee Fall, Bori and Churna.

FOR VISITORS

Fauna:
Tiger, leopard, Wild dog, Gaur, Sambar, Chital, Barking deer, Nilgai, Chowsingha, Giant and Flying squirrel, Fresh water crocodiles and different species of fish.

Flora:
Teak, sal, saja, tendu, bija, haldu, aonla and endemic ferns.

Best Season:
December to April.

SATPURA
TRAVEL GUIDE

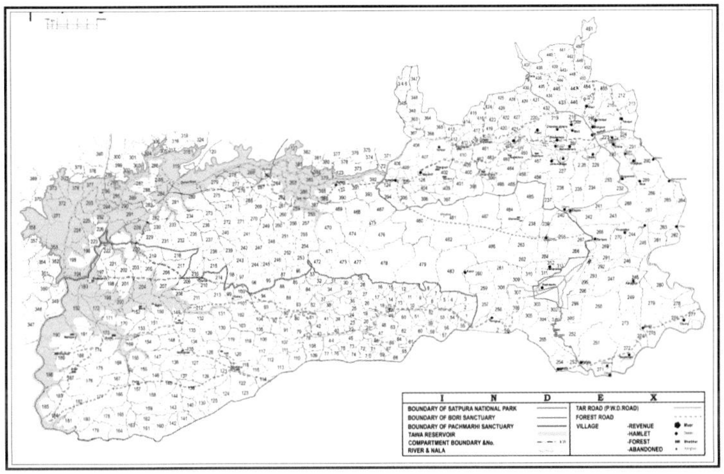

Contact
Field Director
Satpura Tiger Reserve
Hoshangabad - 461001
Madhya Pradesh
Telefax No. (07574) 254394 (O)
E-mail: fd_str@yahoo.com

Core Area	: 1,339.26 sq km
Longitude	: 77° 50" to 78° 30" E
Latitude	: 22° 15" to 22° 45" N
Altitude	: 1,352 m above msl
Rainfall	: 1,200 - 2,000 mm

Temperature

Minimum	: 4° C
Maximum	: 45° C

Seasons

Summer	: March to mid-June
Monsoon	: Mid-June to September
Winter	: November to February

The reserve is closed from the 1st of July to 15th of October for visitors every year (dates are subject to change).

Reaching there
Satpura is accessible from Bhopal (210 km), Jabalpur (240 km) and Nagpur (250 km), which are also the convenient rail/air heads.

Accommodation
Available with MPSTDC, Forest Department, PWD and at private hotels/lodges.

LEFT
The Asiatic Wild dog, commonly known as Dhole, can often be seen on forest paths, river beds, and in forest clearings.

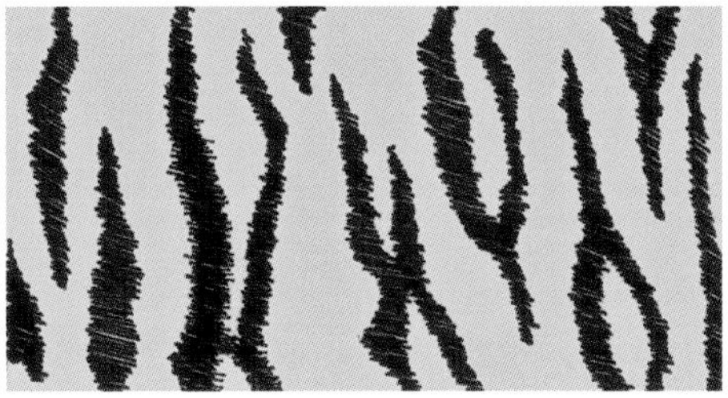

OTHER PROTECTED AREAS

- KUNO-PALPUR
  Wildlife Sanctuary
- RATAPANI
  Wildlife Sanctuary
- MADHAV
  National Park

LEFT

Black partridge, locally known as Kala teetar, is found mainly in the areas which fall under cultivation and scrub.

KUNO-PALPUR WILDLIFE SANCTUARY

Kuno-Palpur Wildlife Sanctuary, with an area of 344.686 sq km, is located in the Sheopur district of Madhya Pradesh, amidst the Vindhaya hill range, close to Rajasthan.

Many streams flow through the protected area, draining into the larger Kuno River, which flows almost through the middle of the park. Kuno is the main source of water for wild animals, since other riverbeds are seasonal.

The forest vegetation comprises of the northern tropical deciduous type with kardhai, salai, dhaora and khair. Gregarious patches of salai, kardhai and khair characterise the Kuno landscape.

◂ LEFT

The Kuno River flows right through the middle of Kuno-Palpur Wildlife Sanctuary.

The wildlife consists of Spotted deer, Sambar, Barking deer, Nilgai, Chinkara, Wild dogs and leopards. There is very little evidence of tiger in the area.

Kuno is also rich in reptiles like the Indian monitor lizard, python, cobra, viper and krait. Various species of fish and Fresh water crocodiles are found in the Kuno River.

The area has a rich avifauna with some migratory birds. The common birds include bayas, egrets, Saras cranes, vultures, hawks and eagles.

Kuno has been selected as an alternate home for the endangered Asiatic lion, which is now confined only to the Gir National Park and Sanctuary of Gujarat. The area is a historical range of Asiatic lions.

The Lion Reintroduction Project is a collaborative initiative between Government of India, state government and the Wildlife Institute of India. A large number of human settlements have been rehabilitated outside the sanctuary area for implementing the lion project. A large number of feral cattle are seen in the area, which are domestic animals abandoned by the villagers during relocation.

▲ ABOVE

Wild pigs, or boars, easily seen in wildlife sanctuaries in Madhya Pradesh, usually move about in groups called sounders. They can be extremely vicious if cornered or taken by surprise.

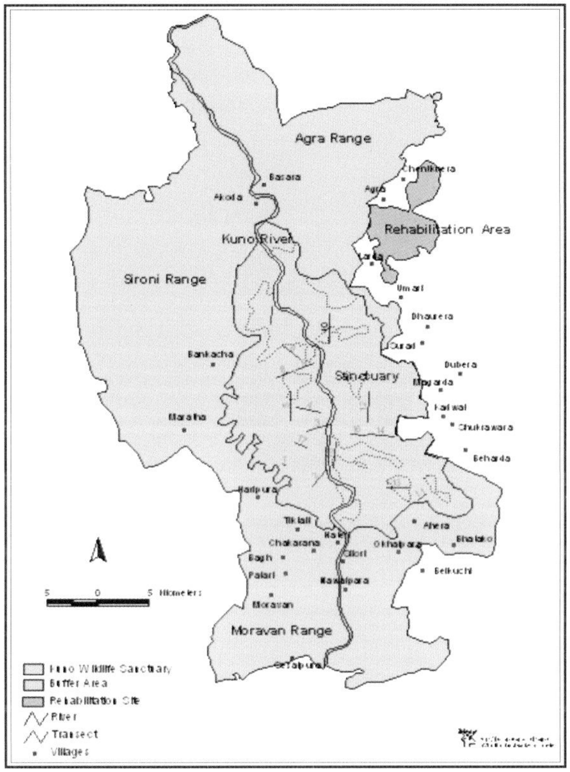

KUNO-PALPUR TRAVEL GUIDE

Core Area	: 344.686 sq km
Longitude	: 77°7' to 77°26' E
Latitude	: 25°20' to 25°53' N
Altitude	: 400 – 600 m above msl
Rainfall	: 1,000 mm

Temperature

Minimum	: 4° C
Maximum	: 45° C

Seasons

Summer	: March to mid-June
Monsoon	: Mid-June to September
Winter	: November to February

Reaching there

Kuno-Palpur is accessible from Gwalior (145 km) by rail, road and air.

Accommodation

Limited accommodation available only with Forest Department.

Contact

Chief Conservator of Forests
Lion Project
Near Tarun Pushkar Central Nursery Campus
Gwalior - 474001
Madhya Pradesh.
Telefax No. (0751) 2235292 (O)
E-mail: cfwildlifegwl@gmail.com

FOR VISITORS

Flora and Fauna

Khardhai, salai, dhaora and khair; Spotted deer, Sambar, Barking deer, Nilgai, Chinkara, Wild dog, leopard, Monitor lizard, python, cobra, viper and krait.

Best season

February to June. The reserve is closed from the 1st of July to 15th of October for visitors every year (dates are subject to change).

RATAPANI
WILDLIFE SANCTUARY

Ratapani is located in the Vindhaya range of Madhya Pradesh, spreading over two revenue districts – Raisen and Sehore. The tract is rich in wildlife and other tourist attractions.

The landscape of Ratapani is undulating with hills, plateau, valleys, plains, gorges and water bodies. Several seasonal streams travel through the sanctuary.

Ratapani is the most important permanent water source for wild animals. Water availability is a major limiting factor for wildlife during the harsh summer months.

The forests comprise of moist and dry teak and mixed forests, along with their associates. The floral species include teak, bija, saja and bhirra. Gregarious patches of khair, bhirra and saja are common in the sanctuary. Bamboo is a common undergrowth in the forest.

The fauna of Ratapani include tiger, panther, Wild dog, hyena, jackal, Chital, Sambar, Nilgai, Sloth bear and Wild pigs. A large number of reptiles are found along with 150 species of birds.

Controlling the biotic pressure due to grazing by cattle apart from poaching and encroachment are the main challenges facing the management.

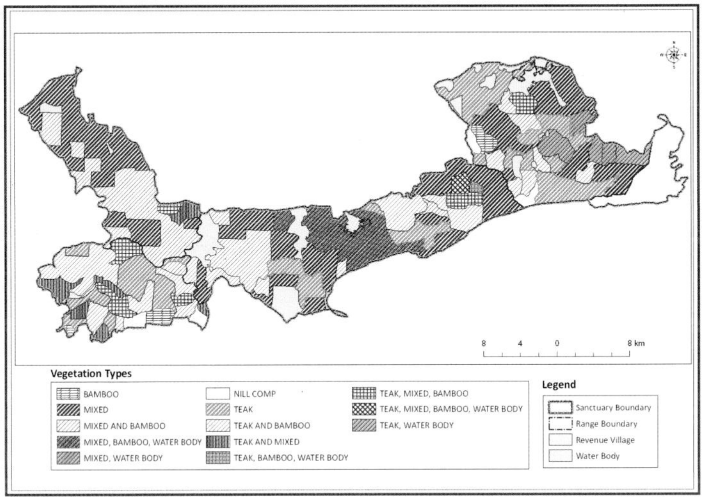

The places of interest include Bhimbetka, Delawari, Ginnorhgarh Fort, Ratapani Dam, Kairi Mahadeo and Kherbana Mandir.

FOR VISITORS

Flora and Fauna

Teak, bija, saja and bhirra; tiger, panther, Wild dog, hyena, jackal, Chital, Sambar, Nilgai, Sloth bear and Wild pigs, apart from 150 species of birds.

Best season

February to June. The reserve is closed from the 1st of July to 15th of October for visitors every year (dates are subject to change).

RATAPANI TRAVEL GUIDE

Core Area	: 906.192 sq km
Longitude	: 77° 31’32” to 78° 4’3”E
Latitude	: 20° 49’7” to 23° 6’17” N
Altitude	: 300 - 690 m above msl
Rainfall	: 900 - 1,000 mm

Temperature

Minimum	: 8° C
Maximum	: 45° C

Seasons

Summer	: March to mid-June
Monsoon	: Mid-June to September
Winter	: November to February

Reaching there

Ratapani is accessible from Bhopal (35 km) by rail, road and air.

Accommodation

Available at the Jungle Camp, Delawadi and at the Highway Treat, Bhimbetka (MPSTDC).

Contact

Divisional Forest Officer
Ratapani Wildlife Sanctuary
Obaidullaganj - 464993
District - Raisen (Madhya Pradesh)
Telephone No. (07480) 224062 (O)
Fax No. (07480) 224478
E-mail: dfo_obaidullaganj@rediffmail.com

MADHAV
NATIONAL PARK

Madhav National Park is located in close proximity to the Shivpuri town. Two National Highways (Agra-Mumbai NH-3 and Shivpuri-Jhansi NH-27) pass through the national park. Madhav National Park has been the hunting preserve of the Maharajas of Gwalior, and is one of the oldest protected areas of Madhya Pradesh.

The Shivpuri area is legendary for its faunal riches. It is said that Emperor Akbar, while returning from Malwa in 1584, captured a large number of wild elephants in its forests. However, at present there are no elephants in this tract.

The *machaan*, or hunting box of the erstwhile rulers still exists, close to the Chand Patha Dam. It has been reported that in 1916, Lord Hardinge shot as many as eight tigers in the area on a single day! During his trip to the Gwalior State, Lord Minto had also shot nineteen tigers in the Shivpuri forests. The area does not have any viable wild population of tigers at present, but a few straying tigers occasionally visit the national park.

◂ LEFT

Water bodies in the reserves are abundant with migratory water birds, a sight not to miss if visiting the protected areas in the state.

During 1956, an area of 165.32 sq km was notified as a national park under the Madhya Bharat National Park Act of 1955. Subsequently, the area has been increased to 354.61 sq km. The extended area is located in the eastern side of the original park, linked by a corridor comprising of revenue and private holdings.

The national park comprises of low Vindhya Hills which are typical of the Malwa highlands and northern tropical dry mixed deciduous forests. The dominant species are kardhai, salai, dhaora and khair. The areas adjoining the Sakhya Sagar Lake provide grasslands for herbivores.

The common fauna in Madhav National Park include leopard, Wild dog, wolf, jackal, hyena, Spotted deer, Nilgai, Chinkara, Chowsingha and Wild pig. There are reports of tigers straying into the area from adjoining forests especially from those close to Ranthambhore.

More than 227 species of birds can be seen, and a large number of migratory birds visit the park during winter – Spot bills, pelicans, Spoon bills, Brahminy ducks and Bar headed geese. The lakes have a large variety of fishes.

▲ ABOVE

A variety of orchid species can be seen across parks in Madhya Pradesh.

The park management faces several challenges like disturbance on account of highways, poaching, illicit mining in the periphery, illicit felling and encroachment, apart from biotic disturbance on account of a large number of villages in the outer areas.

The places of tourist attraction in the area include Sakhya Sagar Lake, Sailing Club and shooting box, Madhav Lake, George Castle, Bhura Kho, Jal Mandir and Siddha Baba.

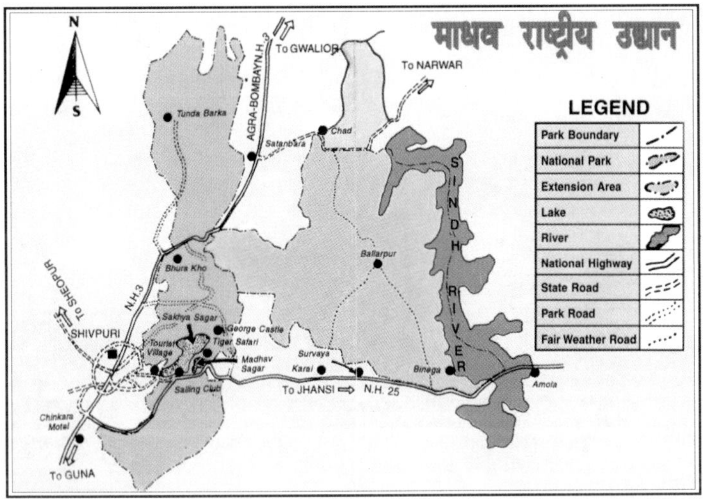

FOR VISITORS

Flora and Fauna

Khardhai, salai, dhaora and khair; leopard, Wild dog, wolf, jackal, hyena, Spotted deer, Nilgai, Chinkara, Chowsingha, Wild pigs and migratory birds.

Best season

The park is open to visitors throughout the year; winter months are suitable for viewing migratory birds.

MADHAV TRAVEL GUIDE

Core Area	: 354.61 sq km
Longitude	: 77°38" to 77°56" E
Latitude	: 25°20'45" to 25°36'36" N
Altitude	: 360 - 480 m above msl
Rainfall	: 1,010 mm

Temperature

Minimum	: 9° C
Maximum	: 43° C

Seasons

Summer	: March to mid-June
Monsoon	: Mid-June to September
Winter	: November to February

Reaching there

Madhav National Park is accessible from Gwalior (108 km) by rail, road and air.

Accommodation

Available at the Tourist Village (MPSTDC), Forest Department and private hotels/ lodges.

Contact

Chief Conservator of Forests & Director
Madhav National Park
Shivpuri-473551
Madhya Pradesh
Telefax No. (07492) 223379 (O)
E-mail: madhavpark@rediffmail.com

PROMINENT WILDLIFE OF MADHYA PRADESH

- BIG CATS
  Tiger and Leopard
- OTHER WILDLIFE

LEFT

Vultures are amongst the most commonly seen birds in Madhya Pradesh, not just in the wild, but in towns and cities as well.

BIG CATS
TIGER AND LEOPARD

The tiger and leopard belong to *Felidae* or the cat family. Only fifteen species of wild cats survive in India. The Indian cheetah, the sixteenth wild cat became extinct in 1948. Including Madhya Pradesh, the tiger is found in seventeen Indian states.

As a taxonomic group, cats exhibit a number of special attributes – size variation, diploid chromosome number of thirty-eight, sexual dimorphism, coat pattern, possessing sweat and sebaceous glands, a typical teeth design with thirty teeth, having powerful canines, a stout skeletal system, retractile claws, adaptation for walking on toes, having a well developed sense of smell and audition and powerful vision even in dim light. They also have the power to maintain equilibrium, have long tactile hair which serves as receptors near the mouth and eyes. These cats have a capacity to predate through stalking and have typical territorial behaviour and reproductive pattern with post birth care of young ones.

◂ LEFT
The 'big cat' on the prowl in Bandhavgarh Tiger Reserve.

The tiger and leopard come under the category of 'big cats' which are capable of 'roaring', limiting 'purring' only during exhalation. Small cats purr during inhalation as well as exhalation. Unlike small cats, big cats restrict their grooming only to activities like licking and rubbing the nose and forepaws. They do not use their forepaws for holding food (except for the Snow leopard) as small cats do.

Since 1972, a focused endeavour, Project Tiger, has been ongoing in the country to save the endangered tiger in designated tiger reserves. Since its inception, Project Tiger has expanded to thirty-seven tiger reserves in 2009 from nine in 1972 – 73.

TIGER

The tiger is the largest of all the big cats, with the body length for males ranging from 275 to 290 cm, and for females around 260 cm. The body weight for males ranges from 160 to 230 kg, and for females from 110 to 160 kg.

Though tigers have evolved in cooler regions, they have adapted themselves well for surviving in warmer temperatures. In Kanha and Bandhavgarh, it is a very common sight to see tigers cooling themselves in shallow water during the summer months, when the temperature can go up to a scorching 48° C.

Communication

Tigers are solitary animals. The only lasting bondage in the life cycle is between mother and cubs up to three years from birth, apart from the brief courtship period lasting for a few days. Communication is achieved through chemical, visual and vocal signals.

It is a very common sight in the tiger reserves of Madhya Pradesh to see tigers 'marking' their territory through a spray of urine mixed with scent. The tiger droppings also bear scent.

Common visual signals include scraping the ground, clawing of trees or rolling on the ground. Some trees on the Kanha-Kisli road bear a large number of claw marks. Through such communication, tigers advertise their presence, status and readiness for mating. Encumbered tigresses give a high pitched, short call to beckon the young ones from their den.

▲ ABOVE

A tigress with cubs is one of the fiercest and most dangerous animals in the world.

Mating and rearing cubs

Tigers mate at any season during a year. In Kanha and Bandhavgarh, mating has been noticed throughout the year, which lasts for a few days (three to eight days). Tiger mating is an aggressive event, with several bouts of copulation during a day. The gestation period is short, ranging from 90 to 100 days. Usually three cubs are seen in a litter, sometimes the litter size goes up to five.

In Pench, the average litter size is four. In Bandhavgarh, some larger litter resulted in the death of weak cubs.

The mother takes care of the cubs for almost three years. During this period, her territory may shrink considerably, and she may not venture far beyond the den. In 1991, a tigress with three cubs had a lethal fight with a leopard in which both the animals succumbed. The orphaned cubs were reared in captivity and later sent to the Lucknow Zoo.

Tiger cubs are blind and helpless after birth. The tigress takes utmost care of her young ones. An encumbered tigress is one of the most dangerous animals to encounter. There have been several instances of locals being lethally mauled by tigresses trailing with cubs in Kanha. Sub-adult tiger cubs are quite adventurous. A forest guard and his follower in Kanha were chased by sub-adult tiger cubs and narrowly escaped being attacked by them.

Often, sub-adult tiger cubs stray out of their dens and are killed by other male tigers. In the Mukki range of Kanha, two female cubs were decapitated by a resident male. On an average, around five to six cubs die during a year due to lethal attacks from larger resident males.

In Kanha, as well as Bandhavgarh, it is very common to see young cubs feasting over a kill along with their mother. Like any young one, tiger cubs are messy eaters. Their association with the mother is a crucial period which teaches them the jungle craft for stalking and killing their prey. In Kanha on several occasions, sub-adult males along with their mother have been seen sharing a kill with a resident male tiger. Tiger cubs leave their mother after 2-1/2 to 3 years. It is commonly observed that a female cub occupies a territory close to its mother. However, male cubs are seldom seen in their natal areas.

The social organisation of tigers mostly depends on the females which breed and contribute to the population. As per the simulation study done using demographic parameters and life history attributes of the tiger population, at least twenty breeding tigresses are required in an inviolate area of 800 - 1,200 sq km. This core area also requires a surrounding buffer for sustaining the dispersing tigers (old ones, young dispersing animals and surplus breeding animals). This situation is essential, considering the poaching pressure, to sustain the tiger dynamics of 'source' and 'sink' populations.

Feeding

An adult tiger requires on an average 3,500 kg of live prey annually. The kill frequency is more in tigresses with cubs. The tiger home range also seasonally varies as per the availability of food. If an area has a low density of prey, a tiger may have to travel more in its range in search of food. It has been reported that tigers may appropriate 10 to 15 per cent of live prey available in an area, which would also be governed by the predation rate of co-predators in the area (leopard, Wild dog). The food spectrum of a tiger is broad, and includes larger to medium sized animals. In Kanha, some male tigers have been seen to selectively predate on Gaur. On one occasion, a resident male tiger in the Mukki range of Kanha was seen eating a huge Gaur which was alive but unable to move owing to the lethal injury caused by the predator to its hamstring muscles.

Tigers also feed on other co-predators like leopards. In Kanha, a tigress along with her cubs waited for almost ten hours below a tree to kill the leopard which was on the tree top! Also in Kanha itself, on several occasions tigers have been seen feeding on other tigers killed during internecine combats.

The life span of a wild tiger may range from eight to ten years. There have been several cases where tigresses have survived for a long time in Kanha, living for almost eleven years (Badimada). However, the average lifespan should be taken as eight years in an area subjected to intense intra-specific competition and other stress conditions.

LEOPARD

The leopard is the commonest of all the Indian roaring cats. It is seen throughout the state, and prominently figures in the frequent man vs. wild animal conflicts. The average total body length of the male leopard is about 215 cm, and of the female around 185 cm. On an average, an adult leopard weighs around 60 – 65 kg, with the females weighing less.

The leopard is a common co-predator in the tiger habitat of Madhya Pradesh. Its habitat ranges from dense forests to open scrub jungles along country side.

Leopards are good climbers and are active at dawn and dusk. Their range of prey overlaps with those of the tiger, but there is a separation in time and space in the use of common habitat to avoid lethal conflicts.

The prey animals of leopard are smaller in size compared to those of a tiger. On several occasions, leopards were seen feeding on young Spotted deer by dragging the kill to low level branches of large trees in Kanha.

Leopards, like tigers, are territorial animals, advertising their presence through droppings and scent markings. However, in a high tiger density area, these advertisements are less pronounced.

RIGHT

Leopards have an advantage over tigers – they can attack their prey from above.

OTHER WILDLIFE

CHITAL

The Chital, or Spotted deer, is the commonest and widely distributed medium sized deer in the state. It is gregarious in nature, feeding on a number of grasses, leaves, flowers and fruits. Chital is a grazer as well as a browser, and is a good coloniser. Its cruising radius from water is not very high and is essentially an animal of 'edges'.

Spotted deer seldom penetrate deep into dense woodlands, tall reeds or steep hills. It thrives very well in an area which is interspersed with woodland, moderate to low field level cover and grass lands of moderate height as seen in Kanha Tiger Reserve. It is an important prey species for the tiger. The Spotted deer attains its best form in the plains of Madhya Pradesh. A well built stag weighs around 85 kg and the height is 90 cm at the shoulder level. The coat is bright rufous fawn which is always profusely flecked with white spots. The older animals, especially the bucks, are more brownish and darker. On the flanks, there are series of spots arranged in longitudinal rows which appear in the form of linear markings. The antlers have three tines – a long brow tine which is at right angles to the beam, apart from two branch tines at the top. The outer tine, which is also the continuation of the beam, is invariably longer.

◀ LEFT

The Chital is one of the most commonly seen member of the deer family in sanctuaries across Madhya Pradesh.

Large groups comprising several hundreds of animals are a common sight in the grasslands of Kanha during the onset of monsoon. They are social without much inter-specific avoidance, and are frequently found in associations with many forest animals, especially the langurs.

They graze in the early hours of the morning and continue till noon. Antler shedding varies in different habitats. Courtship activity is seen throughout the year. In Kanha, intense breeding activities are seen during mid-April to mid-May and from late June till early July. During these periods, large herds of Chital congregate on meadows, with long drawn rutting calls. The rutting activity is largely confined to early hours of the morning and late evenings.

▲ ABOVE

Gharials basking in the sun at Bandhavgarh Tiger Reserve.

The anti-predatory strategy of the Chital is remarkable. In Kanha and Bandhavgarh, the frequent alarm calls of the Chital signal the presence of a tiger or leopard. The smelling as well as hearing capabilities are well developed in Spotted deer. Antler shedding is seen throughout the year, which peaks before the rutting activity.

In Kanha, Spotted deer occur gregariously with the Barasingha, and compete with the latter for food. However, Chital also serve as good buffer species by protecting Barasinghas from becoming targets of predators.

MUNTJAC

Muntjac, or Barking deer is a small deer, inhabiting dense woodlands in dry moist deciduous and semi-evergreen forests. The animal is somewhat territorial, feeding on a variety of leaves and grasses. It prefers a hilly terrain even in dry areas, but requires water at least twice a day.

The antlers of a Barking deer are small, emerging from pedicels in males. The bony, hair covered pedicels extend down on each side of the face as ridges and hence the animal is also called the 'rib-faced deer'. In females, hairy tufts replace the antlers.

The deer weighs around 20 to 23 kg. Older males have a brown coat. The upper canines of the male are well developed for self-defence. The animal is diurnal, feeding on leaves and grasses. Its call resembles the bark of a dog and hence, its name. When alarmed, the deer runs emitting a series of short barks. Barking deer breed throughout the year, the rutting usually takes place in winter. One or two young ones are born at the commencement of the rainy season. Antlers are shed either in May or June.

The Barking deer is a very shy animal and difficult to photograph. However, their sightings are frequent on freshly cut fire lines in Kanha.

BLACK BUCK

The Black buck is a handsome antelope, with the males having a blackish upper spot. Antelopes are different from deer as they have horns instead of antlers. Unlike antlers, horns are permanent and are not shed periodically.

The Black buck prefers large open grassy tracts. It is commonly seen in cultivation sites, causing the loss of village crops. The courtship of the Black buck is typical, with the dominant male maintaining a territory and a harem. The territorial advertisements comprise of repeated defecation on the same spot, with scent marking from ocular glands on nearby grasses.

Sparring matches between males are common as in the case of deer. In Kanha, Black bucks have disappeared owing to fawn predation and practices fostering tall grass for Barasingha.

CHOWSINGA

Chowsinga is the smallest antelope found in India, and is known as the 'four horned antelope', owing to its two pairs of horns. It is found in open grassy plains and gentler slopes, either singly or in pairs.

GAUR

Gaur is known as the Indian bison and is the largest bovid in the world. It is a massive animal, weighing around 900 kg with a chocolate brown-black coat and characteristic 'white stockings'. Though its eating habits are similar to the Sambar, it is more social than the latter, and is often found in large herds. It is a coarse feeder, traversing large distances.

In Kanha, Gaur is seen in the hilly portions of the habitat during the winter and rainy seasons. The Gaur feeds on the leaves of a number of trees, apart from bark. In the Rukhad Sanctuary, large herds of Gaurs are seen debarking young teak trees. The Gaur manipulates a habitat by feeding on relatively coarse grasses.

The herd size varies from five to thirty animals, but smaller groups are common. Solitary bulls, which are dangerous to encounter, are seen after the breeding season. Rutting is seen in Madhya Pradesh during the rainy season, as new calves (golden yellow in colour) are born in the coming monsoon.

SAMBAR

The Sambar is the largest Indian deer, commonly seen in the protected areas and forests of the state. In Kanha, Bandhavgarh, Satpura and other tiger reserves of the state, it is an important prey species for the tiger.

A well grown stag may weigh between 225 - 310 kg. Sambar is a good browser, but at times consumes grass as well.

Sambar can camouflage very well in its surroundings. In the meadows of Kanha, the antlers of a Sambar sitting amidst grasses look like a pair of dried branches.

In Pench, large herds of Sambar in the draw-down areas are not an uncommon sight. Sambar, unlike Spotted deer can cruise for long distances from water sources. It has moderate vision, but the senses of smell and hearing are well developed.

Sambars are seen in small groups, especially in the hilly terrain. In Kanha, they seldom come to the meadows during the monsoon and winter, when fodder and water are plenty in the hills.

RIGHT

The Sambar is primarily a browser that feeds mainly on coarse vegetation.

The rutting period of Sambar is in winter, but may extend up to early summer. Sambar does not give a rutting call like the Spotted deer. The young fawns are born in the span of July to August after a gestation of almost nine months.

BARASINGHA

Barasingha as the name suggests usually has twelve pointed antlers (*bara* means twelve in Hindi). It is a highly endangered deer, and the central Indian sub-species is adapted for thriving in hard ground conditions of Kanha. The resurrection of Barasingha from near extinction is one of the few success stories in the world, owing to concerted holistic efforts under Project Tiger.

The Central Indian Barasingha, once distributed in many districts of the state up to Bastar, is now confined only to Kanha. The population has also been successfully reintroduced in the Supkhar range (Halon Valley) of the reserve, where large herds had been reported to be seen more than hundred years ago.

It is a large sized specialised deer. The height at shoulder level for the stag is around 135 cm weighing up to 180 kg. The antler's average height is up to 75 cm round the curve having a girth of 13 cm at mid beam. The animal has a woolly, brownish yellow coat. The female lacks antlers and is less dark in colour. The stags are maned. The summer coat of both the sexes is lighter and at times may develop light spots. The fawns also have a spotted coat.

Considerable variation is seen in the antler pattern, and at times there are antlers having more than twelve points. The Barasingha is found gregariously with Spotted deer, and there is no inter-specific avoidance. They feed till late in the morning and also in the evening, taking rest in the hot hours. The Barasingha is exclusively a grass eater. Its eyesight and auditory capabilities are moderate; olfaction is acute. The animal has a highly specialised niche and is totally dependent on grasslands.

In Kanha, the Barasingha shed their antlers by May. The re-growth of antlers coincides with the summer season. The monsoon coat is strikingly orange in colour, and groups of Barasinghas in grassy meadows are a pleasant sight during the season.

The young ones are born in October. The courtship is seen during peak winter (December/January). The long drawn rutting bellows of Barasingha are characteristic of the Kanha meadows during these months. The mother takes care of the young ones till they begin their herd life. Barasinghas are predated by tiger. Their fawns fall prey to several carnivores like Wild dogs, jackals and pythons.

▲ ABOVE

Jackals are mainly scavengers and are notorious for their chilling howls.

CHINKARA

Chinkaras are common in drier habitats like Kuno, some parts of Bandhavgarh and several protected areas in Madhya Pradesh. An adult weighs around 20 – 23 kg, measuring almost 65 cm at the shoulder level. Chinkara is a very elegant animal to look at, but very difficult to photograph. The coat colour is light chestnut, which is darker at the junction with the white underside. There is a white streak on either side of the face. The horns have a somewhat 'S' shaped curvature with rings. It is a gregarious animal, found in small groups. The young ones are born in summer as well as autumn.

Chinkaras feed on leaves, fruits and grass and can survive for a long time without water.

STRIPED HYENA

The hyena is a common scavenger found in the forests of Madhya Pradesh. The body is dog-like and the head is large with heavy fore-quarters. There is a thick dorsal crest of long hair distinctly separated from the cream coloured coat.

The colour may show variations – grey, white, buff or tawny. Transverse stripes are present all over the body. The total length of a male is around 150 cm with a height of about 90 cm, weighing 38.5 kg. The female is comparatively smaller in size.

It is commonly seen in open habitats with boulders and ravines, which provide shelter to the animal. It is a nocturnal animal which prefers to feed on the bony carcasses of animals killed by larger predators, using its powerful jaws. The animal mates in winter and the young ones are born during the following summer.

WILD DOG

The Indian Wild dog is known as Dhole. They move in packs, and bring down their prey by chasing them. The pack hunting of the Wild dog is a well organised technique involving disemboweling the prey (feeding while the prey is alive). Wild dogs serve as remarkable dispersers of prey animals in a habitat. Their long predatory chases make the herds run and occupy different parts of a habitat. Wild dogs play an important role in eliminating weak animals from a herd.

The breeding season starts after the monsoon. The parents join the pack after the pups are grown up enough so as to accompany the pack. Around four to six pups are seen in a litter and the young ones are born during January to February.

WOLF

The wolf is a common animal of the open country. It is usually diurnal in habits but at times is nocturnal also. The height of the animal is around 65 - 70 cm, with a length of 90 - 105 cm, weighing about 18 - 27 kg. The animal has long, powerful jaws, the brows are arching and the forehead is elevated. The coat colour is sandy fawn.

Wolves are found in small packs, and feed on a number of small animals, especially the fawns of ungulates. They mainly breed at the end of the rainy season and most of the pups are born in December; three to nine whelps are born in a litter. The life span of the animal is around twelve to fifteen years. Wolves are highly endangered.

Many times, wolves cause man-wildlife interface conflicts. In 1989, a rabid wolf caused loss of several human lives near the Tala range of Bandhavgarh Tiger Reserve.

INDIAN FOX

The fox is small, slim animal with slender limbs. The tail has a distinct black tip and the coat is grayish, which is in striking contrast with the rufous coloured limbs. Head and body together measure 45 to 60 cm and the tail measures 23 to 35 cm. The weight is between 1.8 to 2 kg.

The fox is found throughout the countryside and in open scrub jungles of Madhya Pradesh. Usually the animal remains confined to self dug burrows during the day time and comes out at dusk. It feeds on small animals, reptiles and fruit. Though a solitary hunter, the Indian fox appears to be sociable in nature. It mates in winter and four cubs are usually born in a litter between February and April.

JACKAL

The jackal is closely related to the wolf but it lacks the arching brows and the elevated forehead. The coat colour varies, but generally it is a mixture of black and white with buff near the shoulders, legs and ears. The height is between 38 to 43 cm, the length including head and body is about 60 to 70 cm, tail is 20 to 27 cm and the weight is around 8 to 11 kg.

The jackal is found throughout the state, from humid dense forests to dry open plains. It usually comes out during the dusk and retires by dawn. It has an easily identifiable howl which is long drawn and high pitched.

Jackals are good scavengers, feeding on dead carcasses. Jackals also predate on weak livestock and poultry. At times they resort to fawn lifting. In Kanha, there have been reports of jackals chasing Chital fawns. Jackals are seen in pairs. The animal breeds all round the year. Jackals have suffered due to hunting for their pelt and also for their pest value.

SLOTH BEAR

The Sloth bear is a common omnivore found in the forests of Madhya Pradesh. It has a long black coat with a white, crescent patch on the chest. The spoor of the Sloth bear appears like large, flat footed impressions with claw marks. The Sloth bear has an omnivorous diet, feeding on eggs, fruits, insects and termites.

An encumbered female Sloth bear is dangerous. On several occasions in Kanha, such bears have fought fiercely with tigers. Sloth bears also attack human beings without any provocation. Several forest guards in Kanha have suffered from serious injuries caused by Sloth bears. It is very common to see large deep holes dug by Sloth bears on forest roads in search of termites.

WILD PIG

Wild pigs are very common in and around the forests of the state. They move in large sounders, the number ranging from five to fifty individuals. Wild pigs are omnivorous, feeding on tubers, insects and carrion.

The 'nests' of Wild pigs are very large accumulations of dry leaves, twigs and grasses on the forest flow. They are prolific breeders, with a litter consisting of four to six young ones. Males are ferocious and give a tough fight when encountered with tigers. The dug out patches found commonly on the forest floor confirm the presence of Wild pigs.

MUGGER

The body length of a Mugger is around 3 to 3.5 m on an average. It is difficult to distinguish a Mugger from an Estuarine crocodile in the field, but it must be borne in mind that these two species do not occur simultaneously in an area. The Mugger has a broad snout. There are four sharp, distinct, raised scales in a row behind the head, known as 'post-occipitals'. The back is provided with sixteen or seventeen transverse and six longitudinal series of bony plates (scutes) and these are embedded in the skin. The skin does not have any armouring ventrally. The tail has two series of flattened vertical scales merging near the tail end. The toes are webbed.

The animal is usually olive in colour with black speckling, which are very distinct when the animal is young. Muggers are found in rivers, lakes and other water bodies of Madhya Pradesh (Panna, Kuno, Chambal, Satpura).

Large animals usually bask on river banks and readily slip into water at the slightest alarm; basking crocodiles keep their mouths open which is a method of heat control. Nests (burrows) are usually seen near the border of lakes falling within the range of Mugger. A regular trail from the river side leads to these nests. Muggers hunt in water and the diet usually comprises of fish or any other large animal which can be easily preyed upon.

The Mugger hisses loudly when threatened and also snaps its jaws and lashes its tail. Adults may also give a roaring sound.

During breeding season the scent glands are said to be active and they assist in locating the conspecifics. Mating is usually seen between the middle of January to March and takes place in water. The eggs are laid by the female in nests dug out in the vicinity of river or water body.

The clutch size depends on the size of the female. Usually three to forty or more eggs are present in a clutch; they are white in colour and the shell is hard. Incubation exceeds two months and may continue up to ninety days. The female guards the nest from predators like Monitor lizards, Wild boars and jackals.

GHARIAL

The Gharial is mainly seen in the Chambal River. The species is endangered. It can be easily identified from other species of crocodiles by its long and narrow snout ending in a bulb-like tip. On each side, the jaws are beset with undifferentiated teeth. The adult male has a large 'pot like' cartilaginous mass on the snout tip and hence it is known as Gharial (*ghara* means a pot in Hindi).

The animal is dark olive or brownish olive in colour; underneath it may be white or yellowish white. The young ones are greyish brown with irregular transverse bands on the body and tail. It lives in deep water at river junctions and also in the deep gorges of hilly areas. It is said that the animals spread out with flood waters of the monsoon and return to their perennial areas at the end of the rains. They are aquatic and do not migrate for long distances on land. Usually they bask on river banks.

A bellowing or groaning noise is emitted by the animal when disturbed. The food mainly comprises of fish; sometimes they ingest birds, smaller mammals or turtles. Like other crocodiles they have small stones in their stomach. The mating of gharial takes place in the winter months between December and January. It is believed that the male Gharial uses the *ghara* on the female snout as a leverage for mounting.

Mating takes place in water; the females participate in breeding when they are about 2.5 m in length and males when they reach an age of thirteen to fourteen years (3 m length). The nesting is seen in late March and nests are situated in sand which appears essential for incubation. The nesting season does not vary by more than ten days in any year. The average clutch size is forty and the eggs are white and hard. The incubation period ranges between seventy-two to ninety-two days. Parental care is seen in the form of nest protection, guarding the young ones against predators and taking care of them when they are mature enough to live in water.

▸ RIGHT

The Nilgai is one of the most commonly seen wild animals in central and northern India. Although the male is often called a blue bull (due to its almost indigo-like colour), the females are light brown in colour.

▸ PAGES 85-86

Content with his day of prowling, this tiger in Pench Tiger Reserve gets ready for some well deserved rest.

NILGAI

Nilgai is a large horse-like antelope, measuring almost 140 - 150 cm at the shoulder level (males). The females look different and are smaller in size, without horns. Both the sexes have manes, and males possess a distinct tuft of hair near the throat. Nilgai is frequently seen feeding on crops near villages. Along with Wild pigs, it causes considerable damage to such areas.

Nilgai graze as well as browse, with a diet made up of leaves, fruits, flowers and grasses. The animal is territorial and advertises its presence through repeated defecation at one spot. It is a gregarious animal seen in small herds and breeding takes place throughout the year.

APPENDIX I
BIRD AND BUTTERFLY SPECIES FOUND IN MADHYA PRADESH

BIRDS

S. No.	Common name	Scientific name
1	Lesser florican	*Sypheotides indica*
2	Great Indian bustard	*Ardeotis nigriceps*
3	Shaheen falkan	*Falco peregrinus*
4	King (Red headed) vulture	*Sarcogyps calvus*
5	Sarus crane	*Grus antigone*
6	Paradise flycatcher	*Terpsiphone paradise*
7	Forest owlet	*Athene blewitti*
8	Green munia (avadavat)	*Amandava formosa*
9	Indian vulture	*Gyps indicus*
10	White rumped vulture	*Gyps bengalensis*
11	Painted stork	*Mycteria leucocephala*
12	Malabar pied hornbill	*Anthracoceros coronatus*

BUTTERFLIES

S. No.	Common name	Scientific name
1	Common pierrot	*Castalius rosimon*
2	Danaid eggfly	*Hypolimnas misippus*
3	Short banded sailor	*Neptis columella*
4	Common mime	*Papilio dissimilis, clytia*
5	Crimson rose	*Pachliopta hector*
6	Bamboo tree brown	*Lethe europa*
7	Paintbrush swift	*Baoris farri*
8	Gaudy baron	*Euthalia lubentina*
9	Common banded peacock	*Papilio crino*
10	Orange oakleaf	*Kallima inachus*

APPENDIX II
SCIENTIFIC NAMES OF PROMINENT WILDLIFE

FAUNA

S. No.	English/ local name	Scientific name
1	Tiger	*Panthera tigris*
2	Panther	*Panthera pardus*
3	Wild dog	*Cuon alpinus*
4	Gaur	*Bos gaurus*
5	Barasingha	*Cervus duvauceli*
6	Chital	*Axis axis*
7	Sambar	*Cervus unicolor*
8	Chowsingha	*Tetracerus quadricornis*
9	Nilgai	*Boselaphus tragocamelus*
10	Sloth bear	*Melursus ursinus*
11	Wild boar	*Sus scrofa*
12	Langur	*Presbytis entellus*
13	Wolf	*Canis lupus*
14	Striped hyena	*Hyeana hyeana*
15	Chinkara	*Gazella gazella*
16	Civet	*Viverricula indica*
17	Rhesus monkey	*Macaca mulatta*
18	Porcupine	*Hystrix indica*
19	Paradise flycatcher	*Terpsiphone paradisi*
20	Indian fox	*Vulpes bengalensis*
21	Jackal	*Canis aureus*
22	Black buck	*Antilope cervicapra*
23	Gharial	*Gavialis gangeticus*
24	Mugger	*Crocodylus porosus*
25	Barking deer	*Muntiacus muntjak*

FLORA

S. No.	English/ local name	Scientific name
1	Sal	*Shorea robusta*
2	Saja	*Terminalia tomentosa*
3	Bija	*Pterocarpus marsupium*
4	Jamun	*Syzygium cumini*
5	Mahua	*Madhuca indica*
6	Semal	*Bombax malabaricum*
7	Aonla/Amla	*Emblica officinalis*
8	Tendu	*Diospyros melanoxylon*
9	Dhaman	*Grewia tiliaefolia*
10	Palas	*Butea monosperma*
11	Kusum	*Schleichera oleosa*
12	Arjun	*Terminialia arjuna*
13	Dhaora/Dhawa	*Anogeissus latifolia*
14	Harra	*Terminalia chebula*
15	Bahera	*Terminalia bellerica*
16	Lendia	*Lagerstroemia parviflora*
17	Bamboo	*Dendrocalamus strictus*
18	Teak	*Tectona grandis*
19	Achar	*Buchanania lanzan*
20	Khair	*Acacia catechu*
21	Kardhai	*Anogeissus pendula*
22	Tinsa	*Ougenia oogenensis*
23	Sesham	*Dalberjia latifolia*
24	Salai	*Boswelia serrata*
25	Haldu	*Adena cardifolia*

The heart of
Incredible India

HOTEL LISTINGS

Madhya Pradesh State Tourism Development Corporation Ltd.
For online bookings and latest tariff, please visit us at *www.mptourism.com*

Tariff as on 15.10.2009
Tariff subject to change without prior notice
Taxes extra as applicable
AP: American Plan, CP: Continental Plan
Facilities

Restaurant — Bar — Transport — Conference — Swimming pool

S. No.	Unit	Category	No. of rooms	Tariff (Rs.) Single	Double	Extra Person	Facilities
	AMARKANTAK						
1	Holiday Homes	AC	6	1190	1190	300	Restaurant, Conference
	Tel: (07629) 269416	Aircooled	4	690	690	200	
	E-mail: hhamarkantak@	AC Tent	10	990	990	250	
	mptourism.com	Family Room	2	790	790	-	
	BANDHAVGARH						
2	White Tiger Forest Lodge	AC	26	3290	3890	800	Restaurant, Bar, Transport, Conference, Swimming pool
	Telefax: (07627) 265406, 265366	Aircooled	12	2290	2890	700	
	E-mail: wtfl@mptourism.com			*(AP rates - inclusive of taxes)*			
	BARGI						
3	Maikal Resort (CP)	AC	6	1590	1590	250	Restaurant
	Tel: (0761) 2914577, 09425324211						
	E-mail: mrbargi@mptourism.com						
	BHEDAGHAT						
4	Motel Marble Rocks (CP)	AC	6	1190	1190	175	Restaurant, Conference
	Tel: (0761) 2830424	Aircooled	5	990	990	150	
	E-mail: mmr@mptourism.com						

S. No.	Unit	Category	No. of rooms	Tariff (Rs.) Single	Double	Extra Person	Facilities
	BHIMBETKA						
5	Highway Treat Tel : (07480) 281558 E-mail: bhimbetka@mptourism.com	AC	3	890	890	150	
	BHOPAL						
6	Palash Residency *** (CP) Tel: (0755) 2553006 2553066, 2553076, 3259000 Telefax: (0755) 2577441 E-mail: palash@mptourism.com	AC Suite AC Dlx AC	6 2 29	4990 2790 2290	4990 2990 2490	500 500 500	
7	Hotel Lake View Ashok*** Tel: (0755) 2660090-93 E-mail: hlvashok@sancharnet.in	Standard AC Suite AC	39 4	4500 7000	4500 7000	750 750	
	BIAORA						
8	Tourist Motel (CP) Tel : (07374) 234333 E-mail: biaora@mptourism.com	AC	6	990	990	250	
	CHANDERI						
9	Tana Bana (CP) Tel: (07547) 252222 E-mail: chanderi@mptourism.com	AC Dlx AC Aircooled	1 4 5	990 890 590	990 890 590	150 150 100	
	CHITRAKOOT						
10	Tourist Bungalow Tel: (07670) 265326 E-mail: chitrakoot@mptourism.com	AC Aircooled Dormitory beds	8 9 12	1090 690 125	1090 690 -	150 100 -	
	DATIA						
11	Tourist Motel (CP) Tel: (07522) 238125 E-mail: datia@mptourism.com	AC	4	890	890	150	
	DELAWADI						
12	Jungle Camp Mob. 09300548401 Email: delawadi@mptourism.com	AC Tent (4 Bedded) Tent	7 1	2190 2990 *(AP rates - inclusive of taxes)*	2690 -	700 -	

S. No.	Unit	Category	No. of rooms	Tariff (Rs.) Single	Double	Extra Person	Facilities
	GWALIOR						
13	Tansen Residency*** (CP) Tel: (0751) 2340370 3249000, 4010555,4010666 E-mail: tansen@mptourism.com	AC Dlx Suite AC	12 24	3290 1690	3290 1690	500 300	
	HALALI						
14	Halali Retreat Tel: 0932192848 E-mail: halali@mptourism.com	AC	2	1690	2190	600	
				(AP rates - inclusive of taxes)			
	JABALPUR						
15	Kalchuri Residency*** (CP) Tel: (0761) 2678491/92, 3269000 Telefax: (0761) 2678493 E-mail: kalchuri@mptourism.com	AC	30	1590	1590	300	
	JHABUA						
16	Tourist Motel Tel: (07392) 244668 E-mail: jhabua@mptourism.com	AC	10	990	990	150	
	KANHA						
17	Baghira Log Huts, Kisli Telefax: (07649) 277227, 277211 E-mail: blh@mptourism.com	AC Dlx AC	4 12	3990 3290	4590 3890	890 790	
				(AP rates - inclusive of taxes)			
18	Tourist Hostel, Kisli Tel: (07649) 277310 E-mail: thk@mptourism.com	Dormitory Beds	24	690	-	-	
				(AP rates - inclusive of taxes)			
19	Kanha Safari Lodge, Mukki Tel: (07636) 290715, (07637) 296029 E-mail: ksl@mptourism.com	AC Aircooled	6 6	2590 1890	3190 2490	700 700	
				(AP rates - inclusive of taxes)			
	KATNI						
20	Tourist Motel (CP) Tel: (07622) 262281-82 E-mail: katni@mptourism.com	AC Aircooled	10 4	990 790	1190 990	200 150	

S. No.	Unit	Category	No. of rooms	Tariff (Rs.) Single	Double	Extra Person	Facilities
	KHAJURAHO						
21	Hotel Jhankar (CP) Tel: (07686) 274063, 274194 E-mail: jhankar@mptourism.com	AC	19	1490	1490	200	
22	Hotel Payal (CP) Tel: (07686) 274064, 274076 E-mail: payal@mptourism.com	AC Aircooled	14 11	1490 890	1490 890	200 100	
23	Tourist Village (CP) Tel: (07686) 274062 E-mail: tvkhj@mptourism.com	AC	9	1490	1490	200	
	KHALGHAT						
24	Tourist Motel Tel: (07291) 263564, 263743 E-mail: khalghat@mptourism.com	AC AC Dlx	6 4	990 1190	990 1190	200 300	
	MADLA (PANNA)						
25	Jungle Camp Tel: 09827749975 E-mail: jcmadla@mptourism.com	AC Tent	9	2590 *(AP rates - inclusive of taxes)*	2890	790	
	MAHESHWAR						
26	Narmada Retreat (CP) Tel: (07283) 273455 E-mail: maheshwar@mptourism.com	AC Dlx/Tent AC/Tent Aircooled *(4 bedded family room)*	10 10 3	1590 1290 1290 *(No CP rates on 4 bedded family room)*	1590 1290 -	200 200 -	
	MAIHAR						
27	Hotel Surbahar Tel: (07674) 233362 E-mail:maihar@mptourism.com	AC Dlx AC Aircooled Dormitory Beds	1 5 5 26	1290 890 590 125	1290 890 590 -	200 150 100 -	
	MANDLA						
28	Tourist Motel (CP) Tel: (07642) 260599 E-mail: mandla@mptourism.com	AC Aircooled	4 8	1090 890	1090 890	200 150	
	MANDU						
29	Malwa Resort (CP) Tel: (07292) 263235	AC Aircooled	10 10	1890 1190	1890 1190	200 150	

S. No.	Unit	Category	No. of rooms	Tariff (Rs.) Single	Double	Extra Person	Facilities
30	Malwa Retreat (CP) Tel: (07292) 263221 E-mail: mretreatm@mptourism.com	AC	2	1490	1490	200	
		Aircooled	6	890	890	100	
	NEEMUCH						
31	Tourist Motel Tel: (07423) 280080 Email: neemuch@mptourism.com	AC	6	890	890	150	
	OMKARESHWAR						
32	Narmada Resort (CP) Tel: (07280) 271455 E-mail: omkareshwar@mptourism.com	AC Dlx	3	1590	1590	250	
		AC	8	1190	1190	200	
		Aircooled (budget)	8	790	790	100	
	NOWGAON						
33	Highway Treat (CP) Tel: (07685) 256425 E-mail: nowgaon@mptourism.com	Aircooled	2	890	890	150	
	ORCHHA						
34	Betwa Retreat ***(CP) Tel: (07680) 252618, 252402 E-mail: betwa@mptourism.com	Heritage suite (AC)	1	4990	4990	700	
		AC	14	1690	1690	300	
		AC Tent	10	1290	1290	200	
35	Sheesh Mahal (Heritage Hotel)(CP) Telefax: (07680) 252624 E-mail: smorchha@mptourism.com	Maharaja Suite (AC)	1	4990	4990	700	
		Maharani Suite (AC)	1	3990	3990	300	
		AC	5	1690	1690	200	
		AC Single	1	1190	-	-	
	PACHMARHI						
36	Amaltas Tel: (07578) 252098 E-mail: amaltas@mptourism.com	AC Dlx	5	1890	1890	250	
		AC	5	1590	1590	200	
37	Champak Bungalow Tel: (07578) 285315, 285316 E-mail: champak@mptourism.com	AC Dlx	7	4190	4690	1090	
		AC Tent	5	2790	3290	990	
		AC	7	2790	3290	990	
				(AP rates - inclusive of taxes)			
38	Glen View Tel: (07578) 252533, 252445 E-mail: gview@mptourism.com	AC Dlx	6	4190	4690	1090	
		AC	15	2790	3290	990	
		AC Tent	4	2790	3290	990	
				(AP rates - inclusive of taxes)			

S. No.	Unit	Category	No. of rooms	Tariff (Rs.) Single	Double	Extra Person	Facilities
39	Hilltop Bungalow Tel: (07578) 252846 E-mail: hilltop@mptourism.com	AC	5	1890	1890	250	
40	Hotel Highlands Tel: (07578) 252099,252399 E-mail: highland@mptourism.com	AC Aircooled	20 20	1490 990	1490 990	200 200	
41	Panchvati Tel: (07578) 252096 E-mail: panchvati@mptourism.com	Aircooled (cottages) AC (huts)	5 5	1690 1890	1690 1890	250 250	
42	Rock-End Manor Tel: (07578) 252079 E-mail: rem@mptourism.com	AC Dlx	6	4190 *(AP rates - inclusive of taxes)*	4690	1090	
43	Satpura Retreat Tel: (07578) 252097 E-mail: satpura@mptourism.com	AC Dlx AC Standard	2 4	4190 3590 *(AP rates - inclusive of taxes)*	4690 4290	1090 890	
44	Woodland Bungalow (DIB) Tel : (07578) 252272 E-mail: woodlands@mptourism.com	Aircooled	4	990	990	200	
	PENCH						
45	Kipling's Court Tel: (07695) 232830,232850 E-mail: kcpench@mptourism.com	AC Aircooled Dormitory Beds	15 5 10	3240 2290 700 *(AP rates - inclusive of taxes)*	3890 2890 -	800 690 -	
	PIPARIYA						
46	Tourist Motel Tel: (07576) 222299 E-mail: pipariya@mptourism.com	AC	4	890	890	200	
	ROOKHAD						
47	Highway Treat Tel: (07695) 290130 E-mail: rookhad@mptourism.com	Aircooled	4	490	490	100	

S. No.	Unit	Category	No. of rooms	Tariff (Rs.) Single	Double	Extra Person	Facilities
	SANCHI						
48	Gateway Retreat (CP)	AC Dlx	10	1790	1990	300	
	Tel: (07482) 266723	AC	6	1590	1690	300	
	E-mail: grsanchi@mptourism.com	AC (4 bedded)	2	1990	-	-	
		Annexe Dorm					
		Beds	8	200	-	-	
				(No CP rates on dormitory)			
	SATNA						
49	Hotel Bharhut (CP)	AC Dlx	7	2090	2090	300	
	Tel: (07672) 226071, 223223	AC	6	1290	1290	200	
	E-mail: hbsatna@mptourism.com	Aircooled	9	890	890	150	
		Budget Aircooled	4	490	490	100	
	SHIVPURI						
50	Tourist Village*** (CP)	AC	17	1490	1690	200	
	Tel: (07492) 223760, 221297	AC (4 bedded)	2	1890	-	-	
	E-mail: tvshivpuri@mptourism.com						
	TAWA						
51	Tawa Resort	House Boat	2	3990	3990	-	
	Tel: (07572)290337	AC	6	2190	2690	800	
	E-mail: tawa@mptourism.com			*(AP rates - inclusive of taxes)*			
	UJJAIN						
52	Shipra Residency*** (CP)	AC Suite	4	3990	3990	300	
	Tel: (0734) 2551495-96	AC Dlx	3	1690	1690	200	
	3269000, 2552402	AC	21	1290	1290	150	
	E-mail: shipra@mptourism.com						
53	Hotel Avantika (Yatri Niwas)	AC	4	890	890	150	
	Tel: (0734) 2511398	Aircooled	2	690	690	100	
	E-mail: avantika@mptourism.com	(4 bedded)	6	890	890	-	
		Dorm Beds	32	90	-	-